METHODS IN MOLECULAR BIOLOGY

Series Editor
John M. Walker
School of Life and Medical Sciences
University of Hertfordshire
Hatfield, Hertfordshire, AL10 9AB, UK

For further volumes:
http://www.springer.com/series/7651

Alpha-1 Antitrypsin Deficiency

Methods and Protocols

Edited by

Florie Borel

*Horae Gene Therapy Center, University of Massachusetts
Medical School, Worcester, MA, USA*

Christian Mueller

*Horae Gene Therapy Center, University of Massachusetts
Medical School, Worcester, MA, USA;
Department of Pediatrics, University of Massachusetts
Medical School, Worcester, MA, USA*

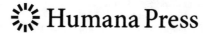 Humana Press

Editors
Florie Borel
Horae Gene Therapy Center
University of Massachusetts
    Medical School
Worcester, MA, USA

Christian Mueller
Horae Gene Therapy Center
University of Massachusetts
    Medical School
Worcester, MA, USA

Department of Pediatrics
University of Massachusetts
    Medical School
Worcester, MA, USA

ISSN 1064-3745   ISSN 1940-6029   (electronic)
Methods in Molecular Biology
ISBN 978-1-4939-8403-9   ISBN 978-1-4939-7163-3   (eBook)
DOI 10.1007/978-1-4939-7163-3

This Humana Press imprint is published by Springer Nature
The registered company is Springer Science+Business Media LLC
The registered company address is: 233 Spring Street, New York, NY 10013, U.S.A.

Preface

Why Did We Write This Book?

The book was triggered by the expansion in the field of alpha-1 antitrypsin (AAT) research. While by no means exhaustive, this book is meant to provide the essential protocols to laboratories ranging from new scientists in the field to more established alpha-one researchers. The aim of this book is to present a solid background by experts in their respective subfield of AAT research as well as state-of-the-art methods that are relevant to the audience, for both clinical and the bench researchers.

Who Is This Book For?

The intended audience of this book is researchers, students, and clinician-scientists interested in AAT deficiency, as well as clinicians in particular those in the fields of pulmonology and hepatology.

What Is in the Book?

This volume of the *Methods in Molecular Biology* series provides a collection of protocols. The book opens with reviews on the pathophysiology of the liver and lungs. Subsequently, protocols are organized in three parts: *Part I* is dedicated to patient-oriented research, *Part II* to animal models, and *Part III* to in vitro work. The book concludes with reviews from experts in the various fields relating to AAT translational research who share their vision of current and/or future therapeutics.

Worcester, MA, USA
Florie Borel
Christian Mueller

Acknowledgments

This book is the result of a long effort from many parties, and we would like to thank the publisher for entrusting us with this task, as well as to convey our appreciation to all our authors for their time and effort on this endeavor. We would like to acknowledge the wonderful alpha-1 patient community who remains a constant source of inspiration to many AAT labs around the world. We would like to dedicate this work to John and Fred Walsh who both have already accomplished so much for the field by increasing awareness and research for patients with alpha-1 antitrypsin deficiency and who are very dear to all the members of the Mueller Lab. Finally, we hope that this volume may find its place in many labs and contribute to the development of new therapeutics for alpha-1 patients.

Contents

Contributors

MARIAM AGHAJAN • *Department of Antisense Drug Discovery, IONIS Pharmaceuticals, Carlsbad, CA, USA*

MYRIAM AOUADI • *Department of Medicine, Karolinska Institutet/AstraZeneca Integrated Cardio Metabolic Center, Karolinska Institutet at Karolinska University Hospital Huddinge, Stockholm, Sweden*

MARCELA APARICIO-VERGARA • *Department of Medicine, Karolinska Institutet/AstraZeneca Integrated Cardio Metabolic Center, Karolinska Institutet at Karolinska University Hospital Huddinge, Stockholm, Sweden*

WILLIAM E. BALCH • *Department of Molecular Medicine, The Skaggs Institute for Chemical Biology, The Scripps Research Institute, La Jolla, CA, USA*

EMELIE BARREBY • *Department of Medicine, Karolinska Institutet/AstraZeneca Integrated Cardio Metabolic Center, Karolinska Institutet at Karolinska University Hospital Huddinge, Stockholm, Sweden*

ERZSÉBET BARTOLÁK-SUKI • *Department of Biomedical Engineering, Boston University, Boston, MA, USA*

IRENE BELMONTE • *Liver Pathology Unit, Departments of Biochemistry and Microbiology, Hospital Universitari Vall d'Hebron, Universitat Autònoma de Barcelona (UAB), Barcelona, Spain; Vall d'Hebron Institut de Reserca (VHIR), Barcelona, Spain*

KEITH S. BLOMENKAMP • *Department of Pediatrics, Saint Louis University, Saint Louis, MO, USA*

FLORIE BOREL • *Horae Gene Therapy Center, University of Massachusetts Medical School, Worcester, MA, USA*

MARION BOUCHECAREILH • *Institut de Biochimie et Génétique Cellulaires, CNRS UMR 5095, Université de Bordeaux, Bordeaux, France*

ANWEN E. BROWN • *UCL Respiratory, University College London, London, UK*

MARTHA CAMPBELL-THOMPSON • *Department of Pathology, Immunology, and Laboratory Medicine, College of Medicine, University of Florida, Gainesville, FL, USA*

MICHAEL CAMPOS • *Division of Pulmonary, Sleep and Critical Care Medicine, Miller School of Medicine, University of Miami, Miami, FL, USA*

ANDREW COX • *Horae Gene Therapy Center, University of Massachusetts Medical School, Worcester, MA, USA*

AIRIEL M. DAVIS • *Cummings School of Veterinary Medicine, Tufts University, North Grafton, MA, USA*

LESLIE J. DONATO • *Department of Laboratory Medicine and Pathology, Mayo Clinic, Rochester, MN, USA*

M. C. ELLIOTT-JELF • *ARUP Institute for Clinical and Experimental Pathology, Salt Lake City, UT, USA*

MAI K. ELMALLAH • *Division of Pulmonary Medicine, Department of Pediatrics, University of Massachusetts Medical School, Worcester, MA, USA; Department of Gene Therapy, University of Massachusetts Medical School, Worcester, MA, USA*

SARAH V. FAULL • *Division of Structural Biology, The Institute of Cancer Research, London, UK*

TERENCE R. FLOTTE • *Horae Gene Therapy Center, University of Massachusetts Medical School, Worcester, MA, USA; Department of Pediatrics, University of Massachusetts Medical School, Worcester, MA, USA; Department of Microbiology and Physiological Systems, University of Massachusetts Medical School, Worcester, MA, USA*

DONGTAO A. FU • *Department of Pathology, Immunology, and Laboratory Medicine, College of Medicine, University of Florida, Gainesville, FL, USA*

ANNE GERSHENSON • *Department of Biochemistry and Molecular Biology, University of Massachusetts Amherst, Amherst, MA, USA*

LILA M. GIERASCH • *Department of Biochemistry and Molecular Biology, University of Massachusetts Amherst, Amherst, MA, USA; Department of Chemistry, University of Massachusetts Amherst, Amherst, MA, USA*

DINA N. GREENE • *Department of Laboratory Medicine, Chemistry Division, University of Washington, Seattle, WA, USA*

DAVID G. GRENACHE • *ARUP Institute for Clinical and Experimental Pathology, Salt Lake City, UT, USA; Department of Pathology, University of Utah School of Medicine, Salt Lake City, UT, USA*

ALISHA M. GRUNTMAN • *Horae Gene Therapy Center, University of Massachusetts Medical School, Worcester, MA, USA*

SHULING GUO • *Department of Antisense Drug Discovery, IONIS Pharmaceuticals, Carlsbad, CA, USA*

NEDIM HADZIC • *Pediatric Centre for Hepatology, Gastroenterology and Nutrition, King's College Hospital, London, UK*

IMRAN HAQ • *UCL Respiratory, University College London, London, UK*

DANIEL N. HEBERT • *Department of Biochemistry and Molecular Biology, University of Massachusetts Amherst, Amherst, MA, USA*

LIZBETH HEDSTROM • *Department of Biology, Brandeis University Waltham, Waltham, MA, USA; Department of Chemistry, Brandeis University Waltham, Waltham, MA, USA*

ANDREW M. HOFFMAN • *Cummings School of Veterinary Medicine, Tufts University, North Grafton, MA, USA*

JAMES A. IRVING • *UCL Respiratory, University College London, London, UK*

MICHAEL KALFOPOULOS • *Division of Pulmonary Medicine, Department of Pediatrics, University of Massachusetts Medical School, Worcester, MA, USA; Department of Gene Therapy, University of Massachusetts Medical School, Worcester, MA, USA*

JOSEPH E. KASERMAN • *Center for Regenerative Medicine (CReM) of Boston University and Boston Medical Center, Boston, MA, USA*

BEENA KRISHNAN • *G.N. Ramachandran Protein Centre, CSIR–Institute of Microbial Technology, Chandigarh, India*

JORGE LASCANO • *Division of Pulmonary, Critical Care and Sleep Medicine, University of Florida, Miami, FL, USA*

ANGELIA D. LOCKETT • *Department of Cellular and Integrative Physiology, Indiana University-Purdue University-Indianapolis, Indianapolis, IN, USA*

STUART MILSTEIN • *Alnylam Pharmaceuticals, Cambridge, MA, USA*

BRETT P. MONIA • *Department of Antisense Drug Discovery, IONIS Pharmaceuticals, Carlsbad, CA, USA*

LUCIANA MONTOTO • *Molecular Biology Department, Hospital de Niños Pedro Elizalde, Buenos Aires, Argentina*

CECILIA MORGANTINI • *Department of Medicine, Karolinska Institutet/AstraZeneca Integrated Cardio Metabolic Center, Karolinska Institutet at Karolinska University Hospital Huddinge, Stockholm, Sweden*

CHRISTIAN MUELLER • *Horae Gene Therapy Center, University of Massachusetts Medical School, Worcester, MA, USA; Department of Pediatrics, University of Massachusetts Medical School, Worcester, MA, USA*

HARIKRISHNAN PARAMESWARAN • *Northeastern University College of Bioengineering, Boston, MA, USA*

KUN QIAN • *Alnylam Pharmaceuticals, Cambridge, MA, USA*

PATRICIA R.M. ROCCO • *Laboratory of Pulmonary Investigation, Carlos Chagas Filho Institute of Biophysics, Federal University of Rio de Janeiro, Rio de Janeiro, Brazil*

FRANCISCO RODRÍGUEZ-FRÍAS • *Liver Pathology Unit, Departments of Biochemistry and Microbiology, Hospital Universitari Vall d'Hebron, Universitat Autònoma de Barcelona (UAB), Barcelona, Spain; CIBER de Enfermedades Hepáticas y Digestivas (CIBERehd), Instituto Nacional de Salud Carlos III, Madrid, Spain*

ALFICA SEHGAL • *Alnylam Pharmaceuticals, Cambridge, MA, USA*

MELISSA R. SNYDER • *Department of Laboratory Medicine and Pathology, Mayo Clinic, Rochester, MN, USA*

BÉLA SUKI • *Department of Biomedical Engineering, Boston University, Boston, MA, USA*

QIUSHI TANG • *Horae Gene Therapy Center, University of Massachusetts Medical School, Worcester, MA, USA; Department of Pediatrics, University of Massachusetts Medical School, Worcester, MA, USA*

JEFFREY H. TECKMAN • *Department of Pediatrics, Saint Louis University School of Medicine, Saint Louis, MO, USA; Department Biochemistry and Molecular Biology, Saint Louis University School of Medicine, Saint Louis, MO, USA; Department of Pediatric Gastroenterology and Hepatology, Cardinal Glennon's Medical Center, Saint Louis, MO, USA*

MICHAELA TENCEROVA • *KMEB, Molecular Endocrinology, University of Southern Denmark, Odense, Denmark*

KRISTEN E. THANE • *Cummings School of Veterinary Medicine, Tufts University, North Grafton, MA, USA*

CHAO WANG • *Department of Molecular Medicine, The Skaggs Institute for Chemical Biology, The Scripps Research Institute, La Jolla, CA, USA*

KAITLYN WETMORE • *Division of Pulmonary Medicine, Department of Pediatrics, University of Massachusetts Medical School, Worcester, MA, USA; Department of Gene Therapy, University of Massachusetts Medical School, Worcester, MA, USA*

ANDREW A. WILSON • *Center for Regenerative Medicine (CReM) of Boston University and Boston Medical Center, Boston, MA, USA*

LINYING ZHANG • *Alnylam Pharmaceuticals, Cambridge, MA, USA*

Chapter 1

Pathophysiology of Alpha-1 Antitrypsin Deficiency Liver Disease

Jeffrey H. Teckman and Keith S. Blomenkamp

Abstract

Classical alpha-1 antitrypsin (a1AT) deficiency is an autosomal recessive disease associated with an increased risk of liver disease in adults and children, and with lung disease in adults (Teckman and Jain, Curr Gastroenterol Rep 16(1):367, 2014). The vast majority of the liver disease is associated with homozygosity for the Z mutant allele, the so-called PIZZ. These homozygous individuals synthesize large quantities of a1AT mutant Z protein in the liver, but the mutant protein folds improperly during biogenesis and approximately 85% of the molecules are retained within the hepatocytes rather than appropriately secreted. The resulting low, or "deficient," serum level leaves the lungs vulnerable to inflammatory injury from uninhibited neutrophil proteases. Most of the mutant Z protein molecules retained within hepatocytes are directed into intracellular proteolysis pathways, but some molecules remain in the endoplasmic reticulum for long periods of time. Some of these molecules adopt an unusual aggregated or "polymerized" conformation (Duvoix et al., Rev Mal Respir 31(10):992–1002, 2014). It is thought that these intracellular polymers trigger a cascade of intracellular injury which can lead to end-organ liver injury including chronic hepatitis, cirrhosis, and hepatocellular carcinoma (Lindblad et al., Hepatology 46(4):1228–1235, 2007). The hepatocytes with the largest accumulations of mutant Z polymers undergo apoptotic death and possibly other death mechanisms. This intracellular death cascade appears to involve ER stress, mitochondrial depolarization, and caspase cleavage, and is possibly linked to autophagy and redox injury. Cells with lesser burdens of mutant Z protein proliferate to maintain the liver cell mass. This chronic cycle of cell death and regeneration activates hepatic stellate cells and initiates the process of hepatic fibrosis. Cirrhosis and hepatocellular carcinoma then result in some patients. Since not all patients with the same homozygous PIZZ genotype develop end-stage disease, it is hypothesized that there is likely to be a strong influence of genetic and environmental modifiers of the injury cascade and of the fibrotic response.

Key words Alpha-1 antitrypsin, Liver, Pathology, Polymers, ER stress, Proliferation, Fibrosis, Cirrhosis, Hepatocellular carcinoma

1 Introduction

The liver is the primary site of synthesis of a1AT protein, although it is also made in enterocytes and some mononuclear white blood cells. The physiologic role of a1AT is to inhibit neutrophil proteases which can leak nonspecifically during the inflammatory response and phagocytosis directed against microorganisms (2). However,

Florie Borel and Christian Mueller (eds.), *Alpha-1 Antitrypsin Deficiency: Methods and Protocols*, Methods in Molecular Biology, vol. 1639, DOI 10.1007/978-1-4939-7163-3_1, © Springer Science+Business Media LLC 2017

recent data has also indicated that a1AT likely has other roles in the immune response. Large quantities of a1AT are secreted from the liver on a daily basis, second only to albumin as mass of a single serum protein. The vast majority of liver disease is associated with homozygosity for the Z mutant of the a1AT gene. During biosynthesis, the a1AT mutant Z protein is appropriately transcribed and translated, and the nascent polypeptide chain is translocated into the ER lumen of the hepatocyte (1, 2). However, unlike the wild-type M, a1AT protein which is rapidly secreted in minutes, the mutant Z form folds inefficiently into its final conformation and is 85% retained in the hepatocyte. Individual, "monomeric" mutant Z molecules are held in the ER, which in experimental systems can last an hour before being directed to proteolysis pathways [4]. However, some of the mutant Z molecules aggregate, or "polymerize," into large masses surrounded by rER. Often, these inclusions, termed globules, are large enough to be seen by light microscopy (Fig. 1). It is this accumulation of a1AT mutant Z protein in hepatocytes which is the inciting event in liver injury associated with a1AT deficiency. While this accumulation is the primary cause of liver damage, it is not sufficient, as not all ZZ individuals develop liver disease despite the presence of mutant Z protein globules in the liver (1) [5]. There is a large role proposed for genetic and environmental disease modifiers, or "second hits."

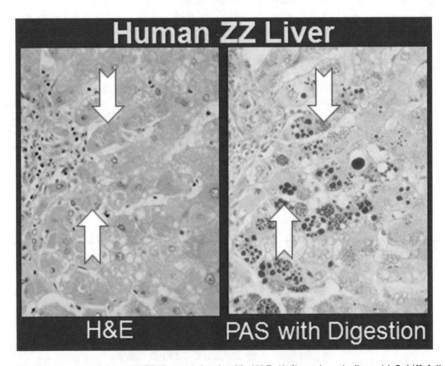

Fig. 1 Photomicrographs of human ZZ liver stained with H&E (*left*) and periodic acid-Schiff followed by diastase digestion (PASd, *right*). PASd stains accumulations of glycoproteins red which can be easily identified on a neutral background. Normal liver is typically free of large, stainable glycoprotein masses. The globules (some highlighted by *arrows*), are variable in size and are not seen in all hepatocytes for unknown reasons

1.1 Molecular and Cellular Pathophysiology of a1AT ZZ Liver Disease

Although accumulation of the mutant Z protein in hepatocytes is the inciting event of liver injury, most of the retained mutant Z protein molecules are eventually directed to intracellular proteolysis pathways and degraded into their constituent amino acids (*see* Fig. 1) [4, 6]. The cell employs a variety of proteolytic processes in an attempt to reduce the intracellular mutant Z protein burden and reduce injury. These include ubiquitin-dependent and ubiquitin-independent proteasomal pathways, as well as other mechanisms sometimes referred to as "ER-associated degradation" (ERAD) [7]. It is thought that the proteasomal pathways as a part of ERAD are the primary route for degradation for a1AT mutant Z monomeric molecules, in the nonpolymerized conformation. Although many of the mechanistic steps in the degradation process, and their specific sequence, are still under investigation, previous work has shown that two molecules present in the ER, calnexin and ER manosidase I (ERmanI), are likely to be critical points of control [7, 8]. Calnexin is a transmembrane ER chaperone which binds a1AT mutant Z, becomes targeted for degradation by linkage to ubiquitin, and then is degraded as this trimolecular complex (a1AT mutant Z–calnexin–ubiquitin) by the proteasome [9]. Studies in human fibroblast cell lines established from ZZ homozygous patients show that patients susceptible to liver disease have less efficient ER-associated degradation of a1AT mutant Z protein than ZZ patients without liver disease [10, 11]. The reduced efficiency of degradation in the liver disease patients presumably leads to a greater steady state burden of mutant Z protein within liver cells and increased liver injury. Studies of the enzyme ERmanI also suggest that it also may have a critical role in directing a1AT mutant Z molecules to the proteasome for degradation. These data raise the possibility that allelic variations in these genes, or in other genes involved in the quality control or proteolytic systems, might alter susceptibility to liver injury by changing the efficiency of degradation. There has been a report that susceptibility to liver disease might also be related to allelic variations in the a1AT gene itself, which would not otherwise be considered disease-associated mutations [12]. Another important proteolytic pathway appears to be autophagy. Autophagy is a highly conserved degradation system in which specialized vacuoles degrade abnormal proteins and larger structures such as senescent organelles. Studies show that the accumulation of the polymerized a1AT mutant Z protein within cells induces an autophagic response, and that autophagy is an important route for the degradation of a1AT mutant Z polymers [6]. In experimental systems liver injury can be reduced by increased autophagic degradation of mutant Z polymer protein [13–15].

Several other lines of evidence support this model of injury from mutant Z protein accumulation. One of the most convincing is studies of the PiZ mouse, which is a transgenic model carrying the human a1AT mutant Z gene, but which retains the wild-type murine anti-protease, a1AT homologous, genes. This model does not develop lung disease, which is thought to be primarily related to a lack of circulating anti-protease activity. However, it does accumulate mutant Z protein in the liver in a way very similar to that seen in human liver, and it does develop a liver disease, very similar in mechanism, histopathology, and clinical course as human ZZ liver disease. Furthermore, studies from several investigators have shown that in cell culture models and in the mouse model, that there is a dose–response relationship between a1AT mutant Z intracellular protein accumulation and cell and liver injury [16]. A number of experimental systems have been used to increase mutant Z protein intracellular accumulation, such as inhibitors of degradation or enhanced inflammation, which increase cell injury markers. Likewise, interventions to reduce mutant Z intracellular accumulation, including siRNA to reduce synthesis and drugs to enhance degradation, have a profoundly positive effect on protecting cells from injury and can even cure the liver injury in the mouse model.

Clinically, liver injury in ZZ humans is usually a slow process which takes place over years to decades, and analysis of human livers have shown that accumulation of a1AT mutant Z protein is very heterogeneous among individual hepatocytes [5, 17]. In the past it has been difficult to reconcile these clinical data with in vitro, cell biological mechanistic studies. New insights from recent studies show that a cellular injury cascade is triggered within the small population of hepatocytes with the largest a1AT mutant Z polymerized protein accumulation. Hepatocytes with the largest a1AT mutant Z protein accumulation, perhaps only a few percent of the total hepatocytes, have increased caspase activation and increased susceptibility to apoptosis [3, 18]. There is also a newly recognized component of oxidative injury [16]. These processes cause a low, but higher than normal baseline rate of hepatocyte death in ZZ liver tissue compared to normal liver. The cells with low polymer accumulation then proliferate to maintain the functional liver mass. Over time, the continued stress, death, and repair leads to liver fibrosis, cirrhosis, HCC, and chronic organ injury. Environmental and genetic modifiers of protein secretion, degradation, apoptosis, or regeneration would then be hypothesized to influence the progression of liver disease in an individual patient [3].

2 Methods/Protocol

Proving that individual hepatocytes with different amounts of mutant Z protein accumulation exhibited variable cell injury and vulnerability to apoptosis was possible by isolating hepatocytes from PiZ mouse liver and assaying the various populations separately, based on mutant Z protein polymer content. Figure 2 is

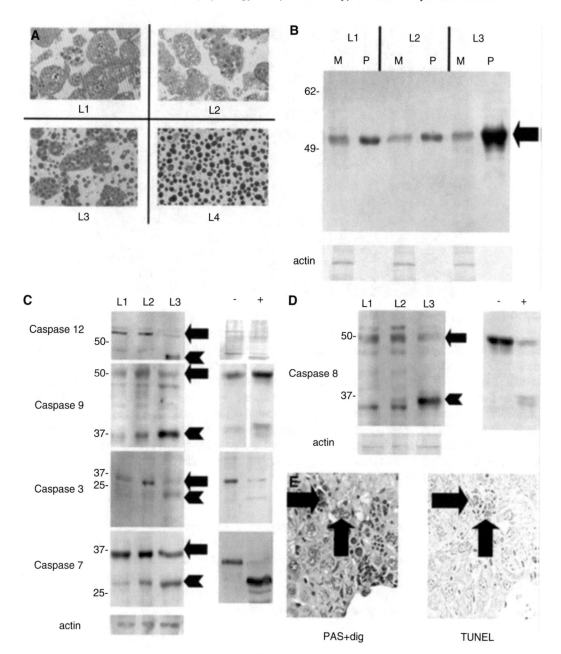

Fig. 2 Analysis of hepatocytes separated on the basis of the a1AT mutant Z protein content with density centrifugation. (**a**) Photomicrographs of PiZ mouse hepatocytes isolated by density gradient centrifugation, as described, in layers L1, L2, L3, and L4. Cells in these layers have progressively more mutant Z protein polymer accumulation and L4 is naked globules. (**b**) Lysates from the intact cells in layers L1, L2, and L3 with the a1AT mutant Z protein monomers (M) separated from the polymers (P), then denatured so the polymers will run at the monomeric molecular weight, and analyzed by sodium dodecyl sulfate–polyacrylamide gel electrophoresis followed by an immunoblot for a1AT. The *arrow* indicates 52 kDa a1AT. The blot for actin as a loading control for each M and P pair is shown (the polymer lane, as previously published, contains only a1AT polymer protein separated from the sample run in the M lane; therefore, there is no actin reactivity in the P lane). (**c**) Immunoblot of layers as shown for caspases. The *arrows* show uncleaved species, and the *arrow heads* show

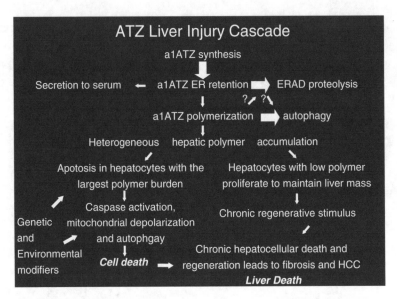

Fig. 3 Liver injury cascade in ZZ, a1AT deficiency liver begins with biosynthesis of the a1AT mutant Z polypeptide chain which enters the ER lumen. There is fold inefficiency and only 15% of the molecules are secreted. The majority of the mutant Z molecules are degraded by ERAD, but some attain the polymerized conformation. Autophagy acts to degrade the protein polymers, but some remain in hepatocytes for long periods. The small population of hepatocytes with the largest mutant Z protein polymers undergoes apoptosis, or death from other mechanisms such as external stressors or redox injury, and the remaining liver cells proliferate to maintain the hepatocellular mass. The cycle of chronic cell death and regeneration activates hepatic stellate cells and other aspects of the fibrotic pathway. The result is liver organ injury, fibrosis, cirrhosis, and hepatocellular carcinoma. All of these steps are likely susceptible to the impact of genetic and environmental modifiers in individual patients, leading to the high clinical variability observed among individuals with the same ZZ genotype

reprinted from the original report describing this phenomenon and which was able, for the first time, to propose a comprehensive model for the intrahepatic injury cascade in a1AT deficiency, which is delineated in Fig. 3 [3]. Similar separation of isolation of hepatocytes based on mutant Z protein polymer globule content is also possible with flow cytometry.

Fig. 2 (continued) cleavage products. When caspases are cleaved they are active as part of apoptotic signal transduction. On the right is shown for each species a negative control (−) of the WT mouse liver whole lysate and a positive control (+) of the PiZ liver lysate pretreated with indomethacin to increase stress and apoptosis, as previously published, for caspases 12 and 9 and pretreated with Jo2 for caspases 3 and 7. (**d**) Immunoblot of layers for caspase 8 with *arrows* as in panel **c**. A WT whole liver lysate for a negative control (−) and a Jo2-treated liver lysate for a positive control (+) are shown on the right. (**e**) Serial sections of liver stained with periodic acid-Schiff followed by digestion to highlight red globules of the a1AT mutant Z protein polymer and TUNEL. The *arrows* show TUNEL-positive cells with associated globules. Molecular weight (relative molecular mass × 103) markers are shown on the left

3 Hepatocyte Isolation

Hepatocyte Isolation and density gradient centrifugation is carried out as described. To minimize bacterial contamination, broad-spectrum antibiotics (GIBCO 15750-060 gentamicin 1:1000, GIBCO 15140-148 penicillin–streptomycin 1:1000) are added to all solutions prior to the start of each experiment. Invitrogen liver perfusion media (LPM) and liver digestion media (LDM) solutions are used as described. Flow cytometry: Cell suspensions are analyzed for scattering and other characteristics using a Beckman flow cytometer. For cell sorting experiments, approximately 1×10^6 cells are suspended in 2–3 mL of PBS. Cells are sorted into four side scatter populations as described in the results and the populations collected into 2 mL of $1 \times$ PBS in 5 mL tubes prelubricated with BSA. Cell fractions are spun at $200 \times g$ 5 min, washed, and the pellet resuspended in Monomer–Polymer buffer for further analysis. For flow cytometric sorting experiments, 5 mL of cell suspension is added to 30 mL of 1.096 g/mL Percoll solution (containing 140 mM NaCl and 10 mM D-glucose in $1 \times$ PBS) and centrifuged for 15 min at 800 RPM in a Beckman Accuspin FR rotor. This produces a single cellular layer which is collected by pipet and washed. Hepatocytes are resuspended in $1 \times$ PBS for flow cytometry. Monomer–Polymer Assay is performed as previously described.

References

1. Teckman JH, Jain A (2014) Advances in alpha-1-antitrypsin deficiency liver disease. Curr Gastroenterol Rep 16(1):367. doi:10.1007/s11894-013-0367-8

2. Duvoix A, Roussel BD, Lomas DA (2014) Molecular pathogenesis of alpha-1-antitrypsin deficiency. Rev Mal Respir 31(10):992–1002. doi:10.1016/j.rmr.2014.03.015

3. Lindblad D, Blomenkamp K, Teckman J (2007) Alpha-1-antitrypsin mutant Z protein content in individual hepatocytes correlates with cell death in a mouse model. Hepatology 46(4):1228–1235. doi:10.1002/hep.21822

4. Teckman JH, Gilmore R, Perlmutter DH (2000) Role of ubiquitin in proteasomal degradation of mutant alpha(1)-antitrypsin Z in the endoplasmic reticulum. Am J Physiol Gastrointest Liver Physiol 278(1):G39–G48

5. Teckman JH, Rosenthal P, Abel R, Bass LM, Michail S, Murray KF, Rudnick DA, Thomas DW, Spino C, Arnon R, Hertel PM, Heubi J, Kamath BM, Karnsakul W, Loomes KM, Magee JC, Molleston JP, Romero R, Shneider BL, Sherker AH, Sokol RJ (2015) Baseline analysis of a young alpha-1-antitrypsin deficiency liver disease cohort reveals frequent portal hypertension. J Pediatr Gastroenterol Nutr 61(1):94–101. doi:10.1097/MPG.0000000000000753

6. Teckman JH, Perlmutter DH (2000) Retention of mutant alpha(1)-antitrypsin Z in endoplasmic reticulum is associated with an autophagic response. Am J Physiol Gastrointest Liver Physiol 279(5):G961–G974

7. Sifers RN (2010) Medicine. Clearing conformational disease. Science 329(5988):154–155. doi:10.1126/science.1192681. 329/5988/154 [pii]

8. Sifers RN (2013) Resurrecting the protein fold for disease intervention. Chem Biol 20(3):298–300. doi:10.1016/j.chembiol.2013.03.002

9. Qu D, Teckman JH, Omura S, Perlmutter DH (1996) Degradation of a mutant secretory protein, alpha1-antitrypsin Z, in the endoplasmic reticulum requires proteasome activity. J Biol Chem 271(37):22791–22795

10. Teckman JH, Perlmutter DH (1996) The endoplasmic reticulum degradation pathway for mutant secretory proteins alpha1-

antitrypsin Z and S is distinct from that for an unassembled membrane protein. J Biol Chem 271(22):13215–13220

11. Perlmutter DH (2011) Alpha-1-antitrypsin deficiency: importance of proteasomal and autophagic degradative pathways in disposal of liver disease-associated protein aggregates. Annu Rev Med 62:333–345. doi:10.1146/annurev-med-042409-151920

12. Sifers RN (2010) Intracellular processing of alpha1-antitrypsin. Proc Am Thorac Soc 7 (6):376–380. doi:10.1513/pats.201001-011AW. 7/6/376 [pii]

13. Pastore N, Blomenkamp K, Annunziata F, Piccolo P, Mithbaokar P, Maria Sepe R, Vetrini F, Palmer D, Ng P, Polishchuk E, Iacobacci S, Polishchuk R, Teckman J, Ballabio A, Brunetti-Pierri N (2013) Gene transfer of master autophagy regulator TFEB results in clearance of toxic protein and correction of hepatic disease in alpha-1-anti-trypsin deficiency. EMBO Mol Med 5 (3):397–412. doi:10.1002/emmm.201202046

14. Kaushal S, Annamali M, Blomenkamp K, Rudnick D, Halloran D, Brunt EM, Teckman JH (2010) Rapamycin reduces intrahepatic alpha-1-antitrypsin mutant Z protein polymers and liver injury in a mouse model. Exp Biol Med (Maywood) 235(6):700–709. doi:10.1258/ebm.2010.009297. 235/6/700 [pii]

15. Hidvegi T, Ewing M, Hale P, Dippold C, Beckett C, Kemp C, Maurice N, Mukherjee A, Goldbach C, Watkins S, Michalopoulos G, Perlmutter DH (2010) An autophagy-enhancing drug promotes degradation of mutant alpha1-antitrypsin Z and reduces hepatic fibrosis. Science 329(5988):229–232. doi:10.1126/science.1190354. science.1190354 [pii]

16. Marcus NY, Blomenkamp K, Ahmad M, Teckman JH (2012) Oxidative stress contributes to liver damage in a murine model of alpha-1-antitrypsin deficiency. Exp Biol Med (Maywood) 237(10):1163–1172. doi:10.1258/ebm.2012.012106

17. Teckman JH (2013) Liver disease in alpha-1 antitrypsin deficiency: current understanding and future therapy. COPD 10(Suppl 1):35–43. doi:10.3109/15412555.2013.765839

18. Teckman JH, Mangalat N (2015) Alpha-1 antitrypsin and liver disease: mechanisms of injury and novel interventions. Expert Rev Gastroenterol Hepatol 9(2):261–268. doi:10.1586/17474124.2014.943187

Chapter 2

Pathophysiology of Alpha-1 Antitrypsin Lung Disease

Michael Kalfopoulos, Kaitlyn Wetmore, and Mai K. ElMallah

Abstract

Alpha-1 antitrypsin deficiency (AATD) is an inherited disorder characterized by low serum levels of alpha-1 antitrypsin (AAT). Loss of AAT disrupts the protease–antiprotease balance in the lungs, allowing proteases, specifically neutrophil elastase, to act uninhibited and destroy lung matrix and alveolar structures. Destruction of these lung structures classically leads to an increased risk of developing emphysema and chronic obstructive pulmonary disease (COPD), especially in individuals with a smoking history. It is estimated that 3.4 million people worldwide have AATD. However, AATD is considered to be significantly underdiagnosed and underrecognized by clinicians. Contributing factors to the diagnostic delay of approximately 5.6 years are: inadequate awareness by healthcare providers, failure to implement recommendations from the American Thoracic Society/European Respiratory Society, and the belief that AATD testing is not warranted. Diagnosis can be attained using qualitative or quantitative laboratory testing. The only FDA approved treatment for AATD is augmentation therapy, although classically symptoms have been treated similarly to those of COPD. Future goals of AATD treatment are to use gene therapy using vector systems to produce therapeutic levels of AAT in the lungs without causing a systemic inflammatory response.

Key words Alpha-1 antitrypsin deficiency, Alpha-1 antitrypsin, Chronic obstructive pulmonary disease (COPD), Emphysema, Neutrophil Elastase

1 Introduction

Alpha-1 antitrypsin deficiency (AATD) is an underrecognized codominantly inherited disorder best known for its physiological effects on the lungs. It was first identified by Laurell and Eriksson approximately 50 years ago when they noted an absence of the alpha-1 band on serum protein electrophoresis [1, 2]. AATD is characterized by reduced amounts of alpha-1 antitrypsin (AAT), a 52-kDa serine protease inhibitor (SERPIN) that is secreted primarily by hepatocytes, although lung and gut epithelial cells, monocytes, and macrophages are secondary sources. AATD is the only genetic risk factor for chronic obstructive pulmonary disease (COPD) and often presents between the ages of 20 and 50 with nonspecific symptoms such as dyspnea, bronchitis, wheezing, and cough [2, 3].

Florie Borel and Christian Mueller (eds.), *Alpha-1 Antitrypsin Deficiency: Methods and Protocols*, Methods in Molecular Biology, vol. 1639, DOI 10.1007/978-1-4939-7163-3_2, © Springer Science+Business Media LLC 2017

MM MS ZZ MM MZ SZ SS

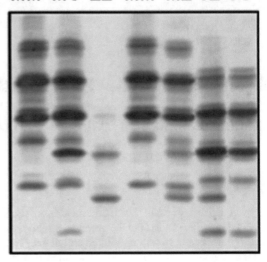

Fig. 1 Qualitative detection of common AAT phenotypes using isoelectric focusing to observe migration rates. Original figure was modified to not include an additional part of image. *Source* [7]

AAT is encoded by the protease inhibitor (PI) locus of the *SERPINA1* gene on chromosome 14q32.1 and is composed of 394 amino acids with its active site at methionine 358. The normal plasma concentration of AAT ranges from 0.9 to 1.75 g/L (15–30 μM/L) with a half-life of 3–5 days. Approximately 123 naturally occurring genetic variants of AAT have been discovered [3–6]. Genetic variants are named depending on speed of migration on an isoelectric focusing (IEF) gel (Fig. 1). For example, the wild-type allele "M" has a medium rate of migration. The two most common mutant alleles of AAT are "Z" (Glut342Lys), accounting for over 90% of disease-causing alleles, and "S" (Glu264Val). The "Z" allele in particular is associated with a severe reduction in plasma AAT levels (to about 5 μM/L) and has the slowest rate of migration on an IEF gel [4, 8, 9].

Due to the many genetic variants of AAT, patients can be classified into four distinct phenotypes: (1) normal: variants of the M phenotype that are prevalent in 99% of the worldwide population and present with normally functioning AAT and plasma levels of ≥20 μM/L; (2) deficient: a mutant phenotype, frequently of the S and Z variants, prevalent in <1% of the worldwide population and associated with plasma AAT levels <20 μM/L; (3) null: a phenotype found in <0.1% of the worldwide population characterized by no detectable AAT in plasma, truncated protein or unstable proteins that are degraded before secretion, and (4) dysfunctional: a phenotype found in <0.1% of the worldwide population characterized by normal amounts of AAT in plasma that do not function correctly and lead to decreased elastase inhibitory activity [6, 10–13]. It is important to note that not all individuals with severe phenotypes such as PI*Z develop disease [14].

2 Epidemiology

Worldwide, it is estimated that more than 3.4 million people have AATD and that there are 116 million carriers of AATD alleles [6]. Affected individuals are commonly white Northern European or of Iberian descent. The incidence of AATD in white newborns is similar to that of cystic fibrosis. Women and men are affected in equal numbers, but men are at greater risk for lung deterioration. Symptomatic individuals have a high mortality rate due to airflow obstruction from lung disease [15–17].

In the USA, it is estimated that 100,000 individuals have AATD, with a prevalence of 1 case per 3000–5000 people [17]. However, only 10% or less of this population is correctly diagnosed [14]. This is supported by an investigation in St. Louis that found only 4% of patients with AATD had been identified and by polls of patients with AATD who reported seeing at least three clinicians before a diagnosis was made [18].

3 Pathophysiology

AAT is composed of three β-sheets and eight or nine α-helices. It expresses a secretion signal that targets it to the endoplasmic reticulum (ER) for folding and glycosylation before it is exported. Proteinase binding cleaves the reactive loop of AAT, which inserts into AAT's β-sheet, trapping the protease and inhibiting it. In a process called "loop sheet polymerization," the tertiary structural change from amino acid mutation results in molecular linkage as the reactive center of one AAT can bind to a gap in the β-sheet of another [19, 20]. The structure of loop sheet polymers of AAT has been elucidated by crystallization [21].

The most common mutation in AAT is the Z mutation, which occurs when there is a substitution of lysine for glutamic acid at position 342. Of the mutated AAT, 15% fold effectively, 15% form polymers, and 70% are degraded by the ER stress response in hepatocytes [1]. In a subset of patients, ER stress and accumulation of AAT in the liver leads to neonatal hepatitis, hepatic cirrhosis, and hepatocellular carcinoma.

After release into the circulation, it is currently believed AAT is internalized into the lungs via clathrin-mediated endocytosis [22]. There, the major function of AAT is to maintain the protease–antiprotease balance. In patients with AATD, the protease–antiprotease balance becomes skewed due to the degradation of mutant AAT by

ER stress, leaving proteases such as matrix metalloproteases, proteinase 3 (PR3), and cathepsins uninhibited [19]. Neutrophil elastase (NE) is the most important protease that AAT inhibits. When left unchecked, NE causes the destruction of lung matrix components, alveolar structures, and blood vessels. Mutant AAT has approximately 5 times less antiproteolytic activity against NE than normal AAT [20]. Furthermore, inhibition of matrix metalloproteinases is important to prevent the degradation of the extracellular matrix. Another protease that AAT inhibits, matriptase, is involved in the activation of epithelial sodium channels. When AAT inactivates the catalytic domain of matriptase, it inhibits the epithelial sodium transport, which results in improved mucociliary clearance in patients with COPD.

AAT plays an essential role in reducing levels of cellular apoptosis in the lungs by inhibition of TNF-alpha, caspase-3, and intracellular cysteine protease. AAT also prevents hemolysis of red blood cells by binding to secreted enteropathogenic *Escherichia coli* proteins, thereby suppressing bacterial proliferation and lung infection in rat models [23].

Another important role of AAT is the attenuation of the neutrophil chemotactic response, a potent cause of inflammation. AAT forms an opposing gradient concentration to interleukin (IL)-8, a ligand for CXCR1 and an important chemokine, so that neutrophils move down an AAT concentration gradient and up an IL-8 gradient. A loss of AAT results in a disrupted concentration gradient and an increased neutrophil chemotactic response. AAT can also bind to soluble immune complexes in order to prevent increased neutrophil chemotactic response [23]. In patients with AATD, unimpeded IL-8 promotes a series of cellular events that contributes to approximately 31% of the neutrophil chemotaxis in the sputum of patients with COPD. Furthermore, alveolar macrophages in patients with AATD release an increased amount of leukotriene (LT) B4. LTB4 is released when uninhibited NE binds to alveolar macrophages and is estimated to contribute to 47% of neutrophil chemotaxis in the sputum of patients with COPD. Patients with a PI*Z phenotype may create AAT polymers that are chemotactic for human neutrophils and cause heightened inflammation by stimulating myeloperoxidase release and neutrophil adhesion, resulting in interstitial neutrophilia [24].

Reactive oxygen species (ROS) produced by neutrophils via the NADPH oxidase enzyme complex are eliminated by AAT through uncertain mechanisms [23]. While ROS is important for killing microbes, release of these free radicals into extracellular space can cause extensive lung damage and inflammation. AAT may prevent asthma and other allergic diseases that cause inflammation by inhibiting IgE-dependent and calcium ionophore-induced histamine release from mast cells [23]. AAT also decreases plasma TNF-alpha levels, induces IL-1 antagonists and inactivates the cytotoxic properties of α-defensins, reducing inflammation reactions [3, 4] Finally, AAT may modify inflammatory signaling pathways such as the cyclic adenosine monophosphate (cAMP)-dependent pathway, responsible for elevating leukocyte production, and inhibit pro-inflammatory cellular signaling [23].

4 Clinical Presentation

For the purpose of this chapter, severe AATD refers to individuals with a form of PI*Z, PI*null and some PI*SZ genotype variants [6]. Severe AATD predisposes patients to COPD, an obstructive disease that leads to early-onset pulmonary emphysema. Emphysema is commonly classified according to forced expiratory volume in 1 s (FEV_1) [13]. It is estimated about 1% of patients with COPD have severe AATD [17]. Individuals with severe AATD are classified as those who have an antitrypsin serum concentration below 35% of mean value or the threshold of 11 μM/L [13, 20]. Patients with AATD most frequently present with dyspnea, but may also present with cough, wheezing, phlegm production, and bronchial hyperresponsiveness [2, 25].

AATD-associated emphysema is predominantly panacinar emphysema and found in the basal region of the lung, compared to more classical COPD emphysema found in the apical region of the lung. A chest CT (Fig. 2) will display this emphysematous change in the base, along with a loss of lung parenchyma and hyperlucency [27]. Emphysema develops as a result of unopposed NE, due to the breakdown of structures necessary for ventilation beyond terminal bronchioles. This leads to enlargement of airspaces and destruction of alveolar walls [28, 29]. Alveolar wall destruction reduces the elastic recoil of the lungs and impairs

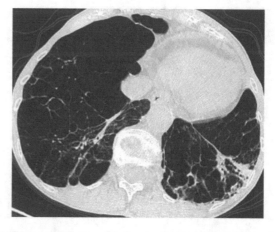

Fig. 2 An AATD associated panlobular emphysema displaying a loss of lung parenchyma. *Source* [26]

their ability to provide traction to small intrapulmonary airways and the pressure needed to inflate bronchi and extrapulmonary airways. Consequently, the airways collapse and obstruction occurs [29, 30].

In a study of patients with severe AATD, most patients had pulmonary function tests that revealed an FEV1 of 50% predicted or lower and a registry of patients with severe AATD had an FEV1 of 43 ± 30% predicted [31, 32]. FEV_1 is a measure of airway obstruction—specifically, it is the measure of the volume of air that can be exhaled in 1 s following a deep inhalation (normal range is 80–120% predicted). PI*ZZ individuals have an annual FEV_1 decline of 23–316 mL. Predictors for accelerated rate of FEV_1 decline include exposure to cigarette smoke, male sex, 30–44 years of age, and an original FEV_1 of between 35% and 79% of expected value [24]. A low FEV_1 is an important risk factor for death in individuals with severe AATD because it implies that patients have severe airway obstruction, progressive lung disease, and emphysema (Fig. 3). Emphysema is the most common cause of death in these individuals, killing approximately 58–72% of the population [34]. Along with a decreased FEV_1, most patients with severe AATD present with an increased functional residual capacity (FRC) [20]. The increased FRC is a consequence of the severe airway obstruction and the subsequent air trapping.

Cigarette smoking is the major risk factor for AATD individuals to develop emphysema [13]. Other risk factors include a history of asthma, chronic bronchitis, and pneumonia. Cigarette smoking is a major risk factor because it contains oxidants that are capable of converting the active site methionine 358 to methionine sulfoxide. This conversion inactivates AAT and reduces its affinity for NE by 2000-fold. In addition, cigarette smoke impairs lung elastase synthesis and recruits inflammatory cells that contribute to the NE

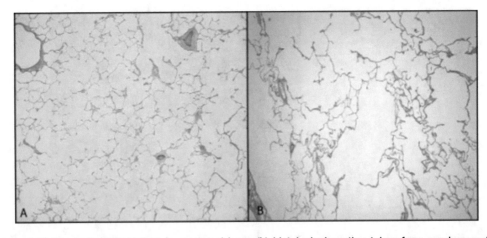

Fig. 3 (**a**) histological section taken from normal lung; (**b**) histological section taken from emphysematous lung. *Source* [33]

load [13]. Smokers and ex-smokers have significantly reduced FEV_1, increasing the risk of airflow obstruction and the development of emphysema [35]. Smokers and ex-smokers who have over 100 cigarettes in their lifetime have worse airway symptoms compared to nonsmokers. In fact, 30% of AATD patients who smoke tend to have significant phlegm and 48% wheeze regularly compared to 13% and 26% of nonsmokers respectively [27]. In addition, Larsson et al. reported that the median onset of dyspnea in smokers with severe AATD was 40 years, compared to 53 years in nonsmokers [36]. The mean life expectancy for smokers with severe AATD is 48–52 years, but this age significantly increases to 60–68 years in nonsmokers [3].

Severe AATD may also lead to bronchiectasis [37]. Bronchiectasis, or the pathological dilation of airways, occurs when the bronchial tree is obstructed due to inflammation, resulting in the destruction of ciliated epithelium by the host immune response. This leads to impaired mucociliary clearance, causing subsequent inflammatory responses and repeat lung infections with persistent inflammation. Over time, bronchi will lose their ability to move air into and out of the lungs as tissue damage extends to the muscle layers [38]. Currently there is debate about whether bronchiectasis is a symptom of AATD because large population based bronchiectasis registries have not shown significant differences between AAT allele frequencies compared to control populations [39]. However, some studies have shown that up to 40% of patients with AATD are affected by bronchiectasis, and in one study by Parr et al. 95% of individuals with severe AATD had a diagnosis of bronchiectasis [27, 37]. A diagnosis of bronchiectasis can be confirmed by a CT scan.

5 Diagnosis

AATD is a widely underdiagnosed and underrecognized disease by clinicians. The diagnostic delay, or time interval between first symptom and initial diagnosis, is estimated to be 5.6 years, with a median time as long as 8 years. Generally younger individuals have shorter diagnostic delays than older individuals [15, 40]. Contributing factors to underrecognition and diagnostic delay are an inadequate awareness of the disease by healthcare providers, a failure to implement evidence-based recommendations and the belief that testing for AATD is not warranted because of the lack of effective and available therapies [14]. It is currently recommended for physicians to test for AATD if any of the following features are found in a patient [13]:

• Emphysema in a patient 45 or younger.

• Emphysema in a nonsmoker or in the absence of a risk factor.

- Emphysema with prominent basilar changes on a chest x-ray.
- A family history of emphysema, bronchiectasis, liver disease or panniculitis.
- Clinical findings of panniculitis or unexplained liver disease.
- Anti-proteinase 3-positive vasculitis.

As previously stated, individuals with severe AATD are classified as those who have an antitrypsin serum concentration below 35% of mean value or the threshold of 11 μM/L with a severe deficient phenotype [20]. Both quantitative and qualitative laboratory testing are typically done to confirm diagnosis of AATD.

Qualitative testing includes radial immunodiffusion and nephelometry, although both tests overestimate serum AAT concentration. Radial immunodiffusion is a technique used to quantitatively estimate antigens and nephelometry is a technique used to determine levels of blood plasma proteins. Threshold levels of 80 mg/dl and 50 mg/dl have been used for radial immunodiffusion and nephelometry, respectively, instead of 11 μM/L [13].

Quantitative testing occurs at the phenotyping and genotyping levels. Phenotyping is done using IEF gels that identify AAT alleles based on their migration patterns. A drawback to phenotyping is it cannot be used to identify those patients with PI*null alleles due to the lack of production of the AAT protein [17]. Genotyping is done by purifying genomic DNA and conducting polymerase chain reaction (PCR), followed by a restriction enzyme digestion and electrophoresis. Melting curve analysis can also be performed after PCR [12, 41].

6   Current Therapy

Treatment of the symptoms of AATD is similar to that of COPD. Pharmacologically, bronchodilators such as long-acting beta-2-agonists with long-acting anticholinergic tiotropium and inhaled corticosteroids can be applied to maximize airflow and provide relief for acute respiratory distress. Antibiotics can be given if a bacterial airway infection that is exacerbating inflammation is suspected [9, 20, 42].

The only FDA approved therapy for AATD is augmentation therapy, or protein replacement by weekly intravenous infusions to restore serum AAT to a therapeutic level of 11 μM/L [4]. This therapeutic level is similar to those found in patients with PI*MZ, a phenotype with little risk for developing significant COPD. Intravenous augmentation therapy has also been shown to reduce mortality and FEV_1 decline [42]. Inhaled therapy is an additional augmentation therapy that is under consideration because it can potentially inhibit airway elastase and reduce elastase-dependent inflammation and damage.

AATD is an attractive target for gene therapy because it a single gene disorder with a relatively short coding sequence and AAT is predominantly found in plasma and extracellular places. Multiple vector systems have been developed to deliver the gene, such as retroviruses, recombinant adenovirus (rAV), and recombinant adeno-associated virus (rAAV) [43]. The most promising of the gene therapies is rAAV, which is capable of inducing therapeutic levels of AAT and is not likely to develop an inflammatory response [44]. Typically, the rAAV vector is delivered via the liver or airway. A novel approach to gene therapy for AATD is to direct small DNA fragments to the liver where they will replace the abnormal DNA sequence of the *SERPINA1* gene [44]. This approach has not yet been tested for efficacy and safety.

Finally, for patients with AATD who progress to end-stage lung disease and therapy is ineffective, lung transplantation (LT) is an option. LT usually only occurs unilaterally due to lack of donors even though a double lung transplant has a better outcome [13]. Another surgical option is lung volume reduction surgery (LVRS). LVRS consists of removing a portion of an emphysematous lung so that the remaining portion can stretch within the thorax. The smaller lung will have more elastic recoil, causing an increase in FEV_1. Additionally, LVRS significantly decreases oxygen consumption and energy expenditure by respiratory muscles because oxygen consumption in emphysematous lungs is increased due to impaired respiratory mechanics [45].

References

1. Gooptu B, Dickens JA, Lomas DA (2014) The molecular and cellular pathology of alpha(1)-antitrypsin deficiency. Trends Mol Med 20 (2):116–127. doi:10.1016/j.molmed.2013.10.007

2. Kessenich CR, Bacher K (2014) Alpha-1 antitrypsin deficiency. Nurse Pract 39(7):12–14. doi:10.1097/01.NPR.0000450385.22603.ce

3. Janciauskiene SM, Bals R, Koczulla R, Vogelmeier C, Kohnlein T, Welte T (2011) The discovery of alpha1-antitrypsin and its role in health and disease. Respir Med 105 (8):1129–1139. doi:10.1016/j.rmed.2011.02.002

4. Flotte TR, Mueller C (2011) Gene therapy for alpha-1 antitrypsin deficiency. Hum Mol Genet 20(R1):R87–R92. doi:10.1093/hmg/ddr156

5. Sinden NJ, Baker MJ, Smith DJ, Kreft JU, Dafforn TR, Stockley RA (2015) Alpha-1-antitrypsin variants and the proteinase/antiproteinase imbalance in chronic obstructive pulmonary disease. Am J Physiol Lung Cell Mol Physiol 308(2):L179–L190. doi:10.1152/ajplung.00179.2014

6. DeMeo DL, Silverman EK (2004) Alpha1-antitrypsin deficiency. 2: genetic aspects of alpha(1)-antitrypsin deficiency: phenotypes and genetic modifiers of emphysema risk. Thorax 59(3):259–264

7. Carroll TP, Floyd O, O'Connor CA, Mcpartlin J, Costello R, O'Neill SJ, Mcelvaney NG (2011) The prevalence of Alpha-1 antitrypsin deficiency in Ireland. Respir Res 12:91

8. Mahadeva R, Lomas DA (1998) Genetics and respiratory disease. 2. Alpha 1-antitrypsin deficiency, cirrhosis and emphysema. Thorax 53 (6):501–505

9. Stockley RA (2014) Alpha1-antitrypsin review. Clin Chest Med 35(1):39–50. doi:10.1016/j.ccm.2013.10.001

10. Stoller JK, Lacbawan FL, Aboussouan LS (1993) Alpha-1 antitrypsin deficiency. In:

Pagon RA, Adam MP, Ardinger HH et al (eds) GeneReviews(R). University of Washington, Seattle, WA

11. Luisetti M, Seersholm N (2004) Alpha1-antitrypsin deficiency. 1: epidemiology of alpha1-antitrypsin deficiency. Thorax 59 (2):164–169

12. Snyder MR, Katzmann JA, Butz ML, Wiley C, Yang P, Dawson DB, Halling KC, Highsmith WE, Thibodeau SN (2006) Diagnosis of alpha-1-antitrypsin deficiency: an algorithm of quantification, genotyping, and phenotyping. Clin Chem 52(12):2236–2242. doi:10.1373/clinchem.2006.072991

13. American Thoracic S, European Respiratory S (2003) American thoracic society/European respiratory society statement: standards for the diagnosis and management of individuals with alpha-1 antitrypsin deficiency. Am J Respir Crit Care Med 168(7):818–900. doi:10.1164/rccm.168.7.818

14. Stoller JK, Brantly M (2013) The challenge of detecting alpha-1 antitrypsin deficiency. COPD 10(Suppl 1):26–34. doi:10.3109/15412555.2013.763782

15. Bornhorst JA, Greene DN, Ashwood ER, Grenache DG (2013) Alpha1-antitrypsin phenotypes and associated serum protein concentrations in a large clinical population. Chest 143(4):1000–1008. doi:10.1378/chest.12-0564

16. Stoller JK, Tomashefski J Jr, Crystal RG, Arroliga A, Strange C, Killian DN, Schluchter MD, Wiedemann HP (2005) Mortality in individuals with severe deficiency of alpha1-antitrypsin: findings from the National Heart, Lung, and Blood Institute registry. Chest 127 (4):1196–1204. doi:10.1378/chest.127.4.1196

17. Silverman EK, Sandhaus RA (2009) Clinical practice. Alpha1-antitrypsin deficiency. N Engl J Med 360(26):2749–2757. doi:10.1056/NEJMcp0900449

18. Silverman EK, Miletich JP, Pierce JA, Sherman LA, Endicott SK, Broze GJ Jr, Campbell EJ (1989) Alpha-1-antitrypsin deficiency. High prevalence in the St. Louis area determined by direct population screening. Am Rev Respir Dis 140(4):961–966. doi:10.1164/ajrccm/140.4.961

19. Brebner JA, Stockley RA (2013) Recent advances in alpha-1-antitrypsin deficiency-related lung disease. Expert Rev Respir Med 7(3):213–229; quiz 230. doi:10.1586/ers.13.20

20. Kohnlein T, Welte T (2008) Alpha-1 antitrypsin deficiency: pathogenesis, clinical presentation, diagnosis, and treatment. Am J Med 121(1):3–9. doi:10.1016/j.amjmed.2007.07.025

21. Huntington JA, Pannu NS, Hazes B, Read RJ, Lomas DA, Carrell RW (1999) A 2.6 a structure of a serpin polymer and implications for conformational disease. J Mol Biol 293 (3):449–455. doi:10.1006/jmbi.1999.3184

22. Sohrab S, Petrusca DN, Lockett AD, Schweitzer KS, Rush NI, Gu Y, Kamocki K, Garrison J, Petrache I (2009) Mechanism of alpha-1 antitrypsin endocytosis by lung endothelium. FASEB J 23(9):3149–3158. doi:10.1096/fj.09-129304

23. Bergin DA, Hurley K, McElvaney NG, Reeves EP (2012) Alpha-1 antitrypsin: a potent anti-inflammatory and potential novel therapeutic agent. Arch Immunol Ther Exp (Warsz) 60 (2):81–97. doi:10.1007/s00005-012-0162-5

24. Stoller JK, Aboussouan LS (2012) A review of alpha1-antitrypsin deficiency. Am J Respir Crit Care Med 185(3):246–259. doi:10.1164/rccm.201108-1428CI

25. Strange C (2013) Airway disease in alpha-1 antitrypsin deficiency. COPD 10(Suppl 1):68–73. doi:10.3109/15412555.2013.764404

26. Sverzellati N, Molinari F, Pirronti T, Bonomo L, Spagnolo P, Zompatori M (2007) New insights on COPD imaging via CT and MRI. Int J Chronic Obstruct Pulm Dis 2 (3):301–312

27. Subramanian DR, Edgar R, Ward H, Parr DG, Stockley RA (2013) Prevalence and radiological outcomes of lung nodules in alpha 1-antitrypsin deficiency. Respir Med 107 (6):863–869. doi:10.1016/j.rmed.2012.12.021

28. D'Armiento J, Dalal SS, Okada Y, Berg RA, Chada K (1992) Collagenase expression in the lungs of transgenic mice causes pulmonary emphysema. Cell 71(6):955–961

29. McElvaney NG, Stoller JK, Buist AS, Prakash UB, Brantly ML, Schluchter MD, Crystal RD (1997) Baseline characteristics of enrollees in the National Heart, lung and blood institute registry of alpha 1-antitrypsin deficiency. Alpha 1-antitrypsin deficiency registry study group. Chest 111(2):394–403

30. Murray JF (1986) The normal lung : the basis for diagnosis and treatment of pulmonary disease, 2nd edn. Saunders, Philadelphia

31. 1994) A registry of patients with severe deficiency of alpha 1-antitrypsin. Design and methods. The Alpha 1-Antitrypsin Deficiency Registry Study Group. Chest 106 (4):1223–1232

32. Evald T, Dirksen A, Keittelmann S, Viskum K, Kok-Jensen A (1990) Decline in pulmonary function in patients with alpha 1-antitrypsin deficiency. Lung 168(Suppl):579–585

33. Yuan R, Nagao T, Paré PD, Hogg JC, Sin DD, Elliott MW, Loy L, Xing L, Kalloger SE, English JC, Mayo JR, Coxson HO (2010) Quantification of lung surface area using computed tomography. Respir Res 11(1):153

34. Brode SK, Ling SC, Chapman KR (2012) Alpha-1 antitrypsin deficiency: a commonly overlooked cause of lung disease. CMAJ 184 (12):1365–1371. doi:10.1503/cmaj.111749

35. Seersholm N, Kok-Jensen A, Dirksen A (1995) Decline in FEV1 among patients with severe hereditary alpha 1-antitrypsin deficiency type PiZ. Am J Respir Crit Care Med 152(6 Pt 1):1922–1925. doi:10.1164/ajrccm.152.6. 8520756

36. Larsson C (1978) Natural history and life expectancy in severe alpha1-antitrypsin deficiency, pi Z. Acta Med Scand 204(5):345–351

37. Parr DG, Guest PG, Reynolds JH, Dowson LJ, Stockley RA (2007) Prevalence and impact of bronchiectasis in alpha1-antitrypsin deficiency. Am J Respir Crit Care Med 176 (12):1215–1221. doi:10.1164/rccm.200703-489OC

38. Stafler P, Carr SB (2010) Non-cystic fibrosis bronchiectasis: its diagnosis and management. Arch Dis Child Educ Pract Ed 95(3):73–82. doi:10.1136/adc.2007.130054

39. Cuvelier A, Muir JF, Hellot MF, Benhamou D, Martin JP, Benichou J, Sesboue R (2000) Distribution of alpha(1)-antitrypsin alleles in patients with bronchiectasis. Chest 117 (2):415–419

40. Stoller JK, Sandhaus RA, Turino G, Dickson R, Rodgers K, Strange C (2005) Delay in diagnosis of alpha1-antitrypsin deficiency: a continuing problem. Chest 128 (4):1989–1994. doi:10.1378/chest.128.4. 1989

41. Andolfatto S, Namour F, Garnier AL, Chabot F, Gueant JL, Aimone-Gastin I (2003) Genomic DNA extraction from small amounts of serum to be used for alpha1-antitrypsin genotype analysis. Eur Respir J 21(2):215–219

42. Stockley RA, Miravitlles M, Vogelmeier C, Alpha One International R (2013) Augmentation therapy for alpha-1 antitrypsin deficiency: towards a personalised approach. Orphanet J Rare Dis 8:149. doi:10.1186/1750-1172-8-149

43. Mueller C, Flotte TR (2013) Gene-based therapy for alpha-1 antitrypsin deficiency. COPD 10(Suppl 1):44–49. doi:10.3109/15412555. 2013.764978

44. Stockley RA, Turner AM (2014) Alpha-1-antitrypsin deficiency: clinical variability, assessment, and treatment. Trends Mol Med 20 (2):105–115. doi:10.1016/j.molmed.2013. 11.006

45. Mora JI, Hadjiliadis D (2008) Lung volume reduction surgery and lung transplantation in chronic obstructive pulmonary disease. Int J Chron Obstruct Pulmon Dis 3(4):629–635

Chapter 3

Measuring and Interpreting Serum AAT Concentration

Leslie J. Donato, Melissa R. Snyder, and Dina N. Greene

Abstract

Deficiency of alpha-1 antitrypsin (AAT) is caused by mutations in the *SERPINA1* gene that results in low concentrations of AAT in circulation. The low AAT concentration can result in uninhibited neutrophil elastase activity in the lung, leading to pulmonary tissue damage and lung disease. Clinical evaluation for possible AAT deficiency includes two critical components: measuring AAT concentration in serum and identification of AAT deficiency alleles. In this chapter the methods by which AAT concentration can be measured in the clinical laboratory are described. The two most common methodologies for AAT quantification employ immunometric techniques, specifically nephelometry and turbidimetry, which are both based on light scatter technology. The AAT in the patient sample is combined with an anti-AAT polyclonal antibody solution leading to polymer formation and a proportional amount of subsequent light scatter. Descriptions of each method are presented, and specifics of quality control and assay parameters are discussed. A special discussion focuses on interpretation of results in the context of the different AAT genetic phenotypes and in the context of patients with active inflammatory conditions. Emerging techniques for AAT quantitation by mass spectrometry are also described given that both AAT quantitation and allele identification can be performed on the same assay.

Key words Alpha-1 antitrypsin, Nephelometry, Turbidimetry, Mass spectrometry

1 AAT Deficiency

In the lungs, there is a fine balance between the proteolytic activity of neutrophil elastase and the inhibitor function of alpha-1 antitrypsin (AAT)—the protease must stay active long enough to function against the invading organism, but must be inactivated before causing excessive damage to the host. Although primarily active in the lungs, AAT is expressed by hepatocytes and, to a much lesser extent, by macrophages and epithelial cells [1, 2]. Hepatically expressed AAT enters the circulation and, because of its relatively small size in comparison to larger protease inhibitors, enters the lungs and other tissues through passive diffusion. Lack of sufficient AAT can lead to lung damage caused by uncontrolled neutrophil elastase activity. Decreased circulating concentrations of AAT may result from liver damage or a specific genetic abnormality. Genetic

Florie Borel and Christian Mueller (eds.), *Alpha-1 Antitrypsin Deficiency: Methods and Protocols*, Methods in Molecular Biology, vol. 1639, DOI 10.1007/978-1-4939-7163-3_3, © Springer Science+Business Media LLC 2017

AAT deficiency is caused by mutations in the SERPINA1 gene that results in the reduced expression (most common) or production of dysfunctional (less common) AAT protein [3].

More than 100 different *SERPINA1* alleles have been documented to date, most of which were originally identified because of altered migration patterns observed by isoelectric focusing (IEF) gel electrophoresis, a technique still commonly used for patient phenotyping. In most ethnic groups, the most common, or wild-type alleles, are designated as "M alleles." In individuals of Northern European descent, the M alleles account for more than 95% of AAT genotypes [4]. The M alleles are classified as nondeficiency alleles because they possess normal protease inhibitory activity and are expressed at normal concentrations. The two most prevalent deficiency alleles that result in lower concentration of functional AAT are the S and Z alleles. Null alleles result in no protein expression at all and subsequent lower total AAT concentrations.

2   Measuring Serum AAT Concentration

The pulmonary pathologies associated with deleterious *SERPINA1* mutations are most commonly caused by decreased AAT concentration. Measurement of AAT at the site of action in the lung would be difficult. Therefore, quantitation of the AAT protein concentration circulating in serum or plasma is used as a surrogate and is a critical component in the assessment of patients suspected of AAT deficiency [5]. Patients homozygous for the Z allele have the lowest serum/plasma AAT concentration and are at the highest risk of developing sequelae from the deficiency. Protein expression from the S allele is reduced, but to a lesser extent as compared to the Z allele; S/S homozygous patients do not usually develop clinical manifestations of deficiency. Patients who are heterozygous for either M/Z or M/S are also usually not at risk of clinical symptoms of deficiency, but may show relatively decreased concentrations of AAT; S/Z compound heterozygotes are considered high risk for developing sequelae. In cases of severe deficiency, AAT protein replacement therapy is the recommended treatment for patients who exhibit pulmonary disease [6].

AAT is an acute phase reactant and therefore the serum concentration results may be elevated during inflammatory states. Consequently, AAT serum concentrations are best evaluated when there is no concurrent inflammation. However, in cases of chronic inflammation or autoimmune disorders it is best to interpret serum AAT concentration in conjunction with phenotype or genotype analysis as a diagnostic tool.

In some cases, apparently low AAT concentration is an incidental finding on a sample submitted for serum protein electrophoresis (SPEP) indicated by a significantly decreased or absent alpha-1 fraction. Although some laboratories will document this abnormality on the report, SPEP should not be relied upon to screen for AAT deficiency as it has limited sensitivity and specificity for detecting small decreases in AAT concentration. Historically, the first tests available to specifically measure AAT protein concentration for the purpose of detecting AAT deficiency were performed by radial immunodiffusion. Such assays have been replaced by more robust immunoassays [7].

Most clinical laboratories quantify AAT using either nephelometry or turbidimetry techniques because they are commercially available, amenable to automation, and precise. Both methods are based on light scatter. After incubating the specimen with anti-AAT polyclonal antibody solution, a polymer matrix between endogenous AAT and the reagent antibodies forms, leading to production of light-scattering large particles. The primary difference between the two methods is the position of the detector relative to the light source. With turbidimetry, the intensity of light transmitted linearly through a sample is measured. As the concentration of light-scattering particles in the sample increases, the intensity of the transmitted light will decrease. In contrast, it is the scattered light that is measured with nephelometry, rather than the transmitted light. In a nephelometer, light is passed through the sample, with the detector being placed at an angle from the sample. When the concentration of particles in the sample increases the intensity of the scattered light, and therefore the amount of light hitting the detector, will increase. For AAT quantification there are no fundamental benefits to either immunoturbidimetry or nephelometry. The decision of which technique a laboratory utilizes is most often a practical decision based on instrument availability, testing volume and overall laboratory organization. Table 1 outlines the differences and similarities between two commonly used methods for AAT quantification, one based on nephelometry and the second on turbidimetry. Example calibration curves and reaction monitors for these methods are illustrated in Figs. 1, 2, 3 and 4.

While not routinely available, quantifying AAT in serum using mass spectroscopy has been described [8]. This method was originally developed as an alternative to gel-based phenotyping in which the presence of the two most common deficiency variants, S and Z, are detected. The assay also has the potential to measure the concentration of AAT when a standard calibration curve is applied to detect a common peptide present in all AAT variants (Fig. 5). In addition, the mass spectrometry assay can also use unique peptides from the different alleles to quantitate individual allele expression in M/Z and M/S patients [9]. Allele-specific quantitation has never been possible using any other method of AAT protein quantitation since all other methods utilize immunometric techniques with antibodies that are designed to recognize all AAT variants. While

Table 1
Comparison of two methods that measure total AAT concentration

Methodology	Turbidimetery	Nephelometry
Instrument (manufacturer)	AU5800 Beckman coulter	BNII Siemens
Specimen type	Serum Heparinized plasma	Serum Heparinized or EDTA plasma
Reference interval	84–218 mg/dL	90–200 mg/dL
Reagents	No preparation of reagents is required. Reagent composition is proprietary, but contains the following active ingredients: TRIS buffer, goat anti-human AAT antiserum, Tween 20, and polyethylene glycol	No preparation of reagents is required. Required reagents include N Antiserum to Human Alpha-1 Antitrypsin, N Diluent, and N Reaction Buffer. In addition, N Protein Standard SL and N/T Protein Controls SL/L, M, and H are needed for calibration and quality control, respectively
Reagent stability	Open reagent bottles are stable for 30 days when stored refrigerated	Open reagent bottles are stable for 4 weeks if stored at 2–8 °C. After being placed on the instrument, reagents are stable for 3 or 5 days (at 8 h/day) for 2 mL or 5 mL bottles, respectively
Analytical measuring range	30–500 mg/dL	1–25 mg/dL (1–1000 mg/dL clinical reportable range)
Specimen volume	0.25 mL minimum	0.50 mL minimum
Calibration	• 5 points • Polygonal calibration; for quantifying unknowns a separate line is drawn between each calibration point, rather than doing a curve fit of all the calibration points. This means there will be five equations of a straight line ($y = ax + b$). • Stable for 14 days	• 6 points • Calibrated using serial dilutions of the N Protein Standard SL in the N Diluent • Diluted standards must be tested within 4 h after dilution • Stable for 3 months unless there is major maintenance or reagent lot change
Reaction type	End point; the optical density is measured initially and at the end point. The difference between these measures is used in calculating the final concentration	Fixed-time kinetic measurement; light-scatter measurements are taken at 7.5 s and 6 min after all reagents are added to the diluted patient sample. The rate of increase in the light scatter is directly proportional to the concentration of AAT in the patient sample

(continued)

Table 1
(continued)

Methodology	Turbidimetery	Nephelometry
Reaction wavelength	600 nm	840 nm
Other	The instrument establishes a reagent blank before the assay is calibrated. The reagent blank is the optical density OD_{600} of all assay components with water added to the reaction instead of serum. The numeric value of this reagent blank is subtracted from all subsequent OD_{600} measurements determined with the reagent (calibration curve, QC material, and patient samples)	Patient samples are initially tested at a 1:20 dilution. If the measurements are either above or below the AAT standard curve, additional dilutions can be performed (clinical reportable range 1–1000 mg/dL)

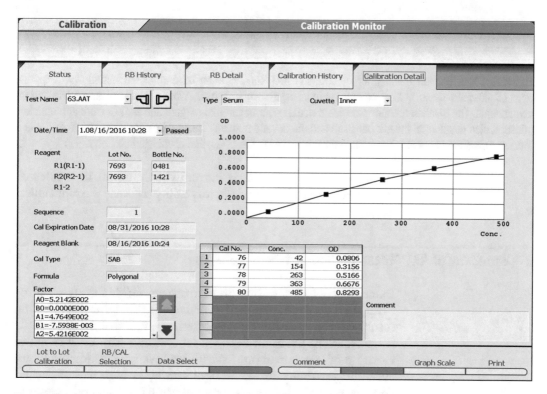

Fig. 1 AU5800 calibration curve for determining AAT concentration by turbidimetry. A polygonal calibration is determined from the five calibration points. For these measurements a separate line is drawn between each calibration point, rather than doing a curve fit of all the calibration points. This means there will be five equations of a straight line ($y = ax + b$). The OD_{600} of the sample and where it fits on the graph will determine which A and B factors you use to calculate the concentration. The calibration factors for each point (A0–A5; B0–B5) are indicated in the table on the *lower left hand side* of the calibration monitor

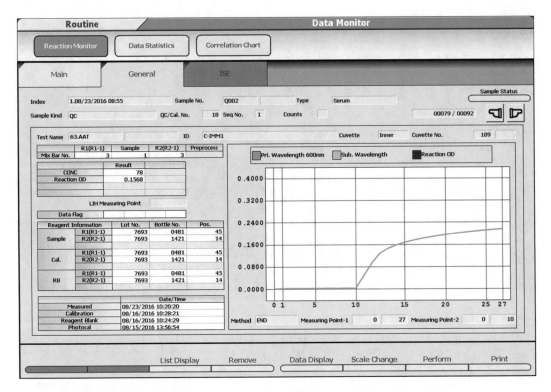

Fig. 2 AU5800 reaction monitor for determining AAT concentration by immunoturbidimetry. An example reaction diagram using quality control material is shown in Fig. 2, which illustrates several important components. The reaction monitor provides a visual confirmation that reagent and sample were appropriately added. It also documents the stir bar used and the cuvette number. These items are important to evaluate if random errors are observed. The result window indicates the OD_{600} and associated concentration

allele-specific quantitation is currently only useful in a research setting, future studies could identify utility that may warrant clinical implementation.

3   Variability of AAT Measurements between Methods

External proficiency testing programs are available through institutions such as the College of American Pathologists (CAP). The results of recent CAP proficiency testing reveal that most methods show excellent precision between laboratories with coefficients of variation of less than about 5% within a given methodology. However, the variability increases significantly, as high as 40%, when comparing mean values across all methods. In general, nephelometric methods report higher concentrations when compared to turbidimetric methods. This finding is likely due to differences in assay calibrations. Internationally certified materials are available to use as primary calibrators for AAT assays (IFCC international reference preparation CRM470 and WHO first IS). However, since

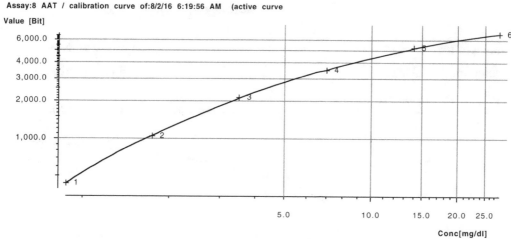

Fig. 3 BNII calibration curve for determining AAT concentration by nephelometry. An example of a 6-point standard curve for AAT on the BNII nephelometer is shown. The standard curve is produced from twofold serial dilutions of the N Protein Standard SL, beginning with a 1:5 dilution. In this example, the standards range from 0.875 mg/dL to 28.0 mg/dL; the values of the standards may vary slightly from one standard lot to another. Calibration acceptability is based on the % deviation, which is a measure of the assigned calibrator value to the calculated value obtained from the mathematically fitted standard curve. For a calibration to be acceptable, the deviation for each point on the curve and the total deviation must be less than or equal to 4.0%

there are no requirements for implementing the reference material as the primary calibrator, no consensus reference method, and no factors to correct for matrix variability (physiological AAT versus purified AAT), AAT assays should not be considered standardized or harmonized.

Most publications consider total AAT concentrations of less than 100 mg/dL as an indication of potential AAT deficiency, but the distribution of results from reference interval studies shows values of approximately 80 mg/dL can define the lower end of the 95th percentile. Given the lack of standardization and observed variability between manufacturer assays, achieving the best clinical sensitivity and specificity relies on laboratories establishing method-specific reference intervals or clinical decision points.

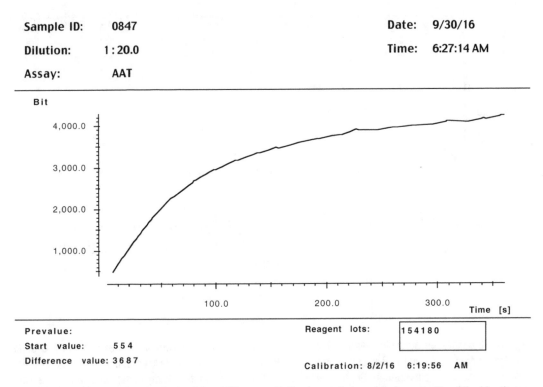

Fig. 4 BNII reaction monitor for determining AAT concentration by nephelometry. An example of the kinetic raw data for the AAT light-scatter reaction on the BNII nephelometer is shown. After all reaction components have been added to the cuvette, the light scatter is measured over time, which is on the X-axis. The amount of scattered light is measured and converted into an electrical signal, which is referred to as a "bit," as seen on the Y-axis. Calculations of AAT concentration use a fixed-time kinetic measurement, which is based on the difference in the bit values at 7.5 s and 6 min

4 Quality Principles for Clinical Laboratory Assays

Proper implementation of FDA cleared automated assays to quantify molecules of interest, such as AAT, requires frequent evaluation of several quality control and assurance measures. Monitoring precision is fundamental to ensuring patient results can be interpreted relative to the laboratory's defined reference interval. To accomplish this, quality control (QC) material, minimally at two different concentrations, should be evaluated with each clinical run. Using commercially available QC material with established peer group means is preferred. Each individual instrument performing the analysis should have a mean and standard deviation established. Rules for acceptability of the QC result should be based on these parameters. The QC material means should aim to monitor the analyte of interest around the medical decision point. For AAT this would be a mean concentration of 70–110 mg/dL.

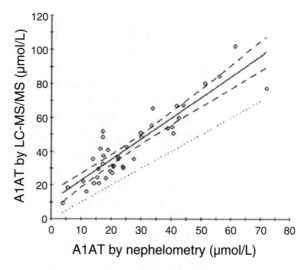

Fig. 5 AAT quantification is possible using mass spectrometry. A comparison of total AAT quantitation by nephelometry and LC-MS/MS is shown. AAT was quantitated in 40 serum samples by nephelometry and by LC-MS/MS using a tryptic peptide common to all AAT alleles. A perfect correlation is indicated by the *dashed line*; the regression analysis of the actual comparison (*solid line*) shows a correlation coefficient of 0.91 (figure reproduced with permission)

Maintaining quality laboratory also requires evaluating accuracy. Due to incommutability of results, accuracy is relative to the instrument. Within the instrument, accuracy is measured using proficiency testing or alternative assessment. Here, an unknown sample is analyzed and compared to a peer group of labs using the identical instrument/assay (proficiency testing) or compared to an individual outside lab evaluating the same sample (alternative assessment). CAP has proficiency testing material for purchase, which is generally the preferred accuracy assessment for the clinical laboratory. Accuracy should be assessed twice per year. Acceptable criteria should be established and samples exceeding those thresholds must be investigated. Accuracy assessment is also accomplished by verifying the analytical measuring range (AMR), which is typically achieved by serially diluting a high sample (into an appropriate diluent or low concentration sample) and calculating percent recovery. Alternatively, commercially available standards can be purchased and used to compare the measured to the expected values.

5   Concentration of AAT in Different Phenotypes

There are over 100 *SERPINA1* alleles documented leading to many possible phenotypes. Understanding the most commonly observed phenotypes and their associated AAT concentrations can facilitate clinical interpretation. There are two published studies documenting the observed AAT concentrations and concurrent phenotype for samples received at national reference laboratories [10, 11]. These studies describe the frequency of phenotype groups in clinically ordered samples, but not in the general population. The frequency of disease alleles is likely higher in these cohorts compared to the general population due to "referral bias." Similarly, while these studies provide valuable information by reporting the expected ranges of serum AAT protein concentrations in the various phenotype groups, no exclusion criteria were applied to the analyses. Since AAT is an acute phase reactant and patients were not screened for inflammation, the upper limits are likely higher than that would be observed in a healthy donor population. However, the lower limit is of clinical importance when evaluating AAT deficiency. As reported in their studies, approximately 80% of all samples sent for AAT evaluation were classified as M/M. The lower limit of the central 95th percentile in pediatrics (0–18 years) with the M/M phenotype was slightly lower compared to adults (≥18 years) (100 mg/dL for adults, and 93 mg/dL for pediatrics). In addition, AAT expression increased marginally with age (lower limit of 90 mg/dL for those age < 1 year to 113 mg/dL in those >80 years) and was slightly higher in women compared to men. These minimal differences, however, are likely not clinically significant, although they are interesting from an inflammatory perspective. No differences in expression were observed between blacks, whites, and Hispanics. However, AAT expression was found to be lower in individuals of Asian descent (lower limit of 82 mg/dL). For non-M/M phenotype groups, the lower limit of the central 95th percentile in adults decreased according to the severity of the deficiency allele: MM (100 mg/dL), M/S (84 mg/dL), M/Z (61 mg/dL), S/S (35 mg/dL), S/Z (35 mg/dL), and Z/Z (12 mg/dL).

6   Using AAT Concentration to Identify Potential Deficiencies

The concentration of AAT can be used to identify patients who may be at the risk of harboring a deficiency allele, and therefore an algorithmic approach using a combination AAT serum concentration and genotyping results is often used to identify deficiency allele-carrying individuals [12, 13]. These algorithms correlate the AAT concentration with the presence or absence of the S and Z alleles detected using allele specific genotyping. Using rules based on expected AAT concentrations for various phenotypes, discordant results are reflexed to an IEF-based phenotyping assay. For example, if a quantitative result of 72 mg/dL is obtained but no S or Z alleles are identified by targeted genotyping, phenotyping would be performed to rule in/out a rare deficiency allele. By using the results of the two, or sometimes three, tests, laboratory directors and physicians can more accurately and efficiently interpret AAT results. The disadvantage of this approach is that at least two tests are run on every patient independent of AAT concentration. Alternatively, it may be possible to use the AAT concentration alone to identify samples that are likely to harbor clinically significant deficiency alleles [11]. Using receiver operator characteristic curve analysis a cutoff of $\leq$60 mg/dL, which has been reported to be the threshold of deficiency symptom onset [14–16], 98.5% of Z/Z individuals are detected with 99.8% specificity, but only 65% of all at-risk individuals are detected ("at-risk" defined as two deficiency alleles including, but not limited to Z/Z; ex: S/S; S/Z). Raising the cutoff to $\leq$85 mg/dL increases the sensitivity for Z/Z detection to 99.5% and all at-risk individuals to 85.9%. To identify all Z/Z individuals a cutoff of $\leq$120 mg/dL would be required. At that cutoff 96.9% of all at-risk phenotypes would be also be detected, but specificity would be very poor. Using a cutoff that maximizes the sensitivity of deficiency allele identification sacrifices specificity in order to identify nearly all at-risk individuals. As with any test, there is a fine balance between clinical sensitivity and specificity. Since total AAT is a screen, cutoff values to reflex the specimen for further testing should optimize sensitivity. Very low concentrations of AAT ($\leq$60 mg/dL) can be considered diagnostic for an at-risk homozygous phenotype.

7   Summary

Quantifying serum/plasma AAT concentration is critical for the assessment of patients with suspected AAT deficiency. Patients with extremely low circulating protein concentrations are at risk for developing clinical manifestations of deficiency. Identification of patients with AAT deficiency is essential to guiding clinical management and preventing damage to the lungs. Immunoassay methods, nephelometry or turbidimetry, are used most commonly in clinical practice because of their precise measuring capabilities and ease of use. However, these quantitation methods are not standardized resulting in variability of results across methodologies. In the future, emerging techniques for AAT quantitation such as mass spectrometry could become more mainstream and may offer novel insights into the pathophysiology of AAT deficiency.

References

1. DeMeo DL, Silverman EK (2004) Alpha1-antitrypsin deficiency. 2: genetic aspects of alpha(1)-antitrypsin deficiency: phenotypes and genetic modifiers of emphysema risk. Thorax 59:259–264

2. Stoller JK, Aboussouan LS (2005) Alpha1-antitrypsin deficiency. Lancet 365:2225–2236

3. Salahuddin P (2010) Genetic variants of alpha1-antitrypsin. Curr Protein Pept Sci 11:101–117

4. Dykes DD, Miller SA, Polesky HF (1984) Distribution of alpha 1-antitrypsin variants in a us white population. Hum Hered 34:308–310

5. Aboussouan LS, Stoller JK (2009) Detection of alpha-1 antitrypsin deficiency: a review. Respir Med 103:335–341

6. Sandhaus RA, Turino G, Stocks J, Strange C, Trapnell BC, Silverman EK et al (2008) Alpha1-antitrypsin augmentation therapy for pi*mz heterozygotes: a cautionary note. Chest 134:831–834

7. Viedma JA, de la Iglesia A, Parera M, Lopez MT (1986) A new automated turbidimetric immunoassay for quantifying alpha 1-antitrypsin in serum. Clin Chem 32:1020–1022

8. Chen Y, Snyder MR, Zhu Y, Tostrud LJ, Benson LM, Katzmann JA et al (2011) Simultaneous phenotyping and quantification of alpha-1-antitrypsin by liquid chromatography-tandem mass spectrometry. Clin Chem 57 (8):1161

9. Donato LJ, Karras RM, Katzmann JA, Murray DL, Snyder MR (2015) Quantitation of circulating wild-type alpha-1-antitrypsin in heterozygous carriers of the s and z deficiency alleles. Respir Res 16:96

10. Donato LJ, Jenkins SM, Smith C, Katzmann JA, Snyder MR (2012) Reference and interpretive ranges for alpha(1)-antitrypsin quantitation by phenotype in adult and pediatric populations. Am J Clin Pathol 138:398–405

11. Bornhorst JA, Greene DN, Ashwood ER, Grenache DG (2013) Alpha1-antitrypsin phenotypes and associated serum protein concentrations in a large clinical population. Chest 143:1000–1008

12. Snyder MR, Katzmann JA, Butz ML, Wiley C, Yang P, Dawson DB et al (2006) Diagnosis of alpha-1-antitrypsin deficiency: an algorithm of quantification, genotyping, and phenotyping. Clin Chem 52:2236–2242

13. Bornhorst JA, Procter M, Meadows C, Ashwood ER, Mao R (2007) Evaluation of an integrative diagnostic algorithm for the identification of people at risk for alpha1-antitrypsin deficiency. Am J Clin Pathol 128:482–490

14. Crystal RG (1989) The alpha 1-antitrypsin gene and its deficiency states. Trends Genet 5:411–417

15. Stoller JK, Aboussouan LS (2012) A review of alpha1-antitrypsin deficiency. Am J Respir Crit Care Med 185:246–259

16. Turino GM, Barker AF, Brantly ML, Cohen AB, Connelly RP, Crystal RG et al (1996) Clinical features of individuals with pi*sz phenotype of alpha 1-antitrypsin deficiency. Alpha 1-antitrypsin deficiency registry study group. Am J Respir Crit Care Med 154:1718–1725

Chapter 4

AAT Phenotype Identification by Isoelectric Focusing

Dina N. Greene, M.C. Elliott-Jelf, and David G. Grenache

Abstract

Isoelectric focusing (IEF) electrophoresis is considered to be the gold standard test for determining an individual's AAT phenotype. IEF electrophoresis is a technique used to separate proteins by differences in their isoelectric point (pI). Testing is performed on serum that is applied to an agarose gel containing ampholytes which create a pH gradient ranging from 4.2 to 4.9. Variants of AAT are therefore separated from each other and, after visualization of the focused protein bands using immunochemical techniques, can be identified and an AAT phenotype determined.

In this chapter we elaborate on IEF electrophoresis as it relates to AAT phenotyping, describe practical approaches to AAT variant identification, and discuss circumstances in which phenotype testing may be inaccurate.

Key words Isoelectric focusing electrophoresis, Immunofixation, Immunodetection, Alpha-1 antitrypsin phenotyping, Test interpretation

1   Introduction

The use of IEF electrophoresis to determine an AAT phenotype is a three-step process. First, serum proteins are isoelectrofocused in an agarose gel to separate serum proteins by their pI. Second, visualization of AAT proteins is accomplished using enzyme-labeled anti-AAT antiserum. Third, the AAT banding pattern of the serum sample is interpreted by an individual who is experienced in identifying AAT variants by this technique.

IEF electrophoresis is ideally suited for AAT phenotyping due to its ability to "focus" proteins at a location in the gel where the pH of the gel matches the pI of the AAT variant. At this location, the net charge of the protein is zero and it ceases to migrate. Any movement of the protein due to ordinary diffusion is counteracted when the protein gains a charge and subsequently migrates back to its pI position. IEF electrophoresis is a high-resolution method because a specific pH region in the gel is very narrow allowing separation of proteins whose pI values differ by only 0.02 pH units [1].

Florie Borel and Christian Mueller (eds.), *Alpha-1 Antitrypsin Deficiency: Methods and Protocols*, Methods in Molecular Biology, vol. 1639, DOI 10.1007/978-1-4939-7163-3_4, © Springer Science+Business Media LLC 2017

The pH gradient in the agarose gel is created by amphoteric polyaminocarboxylic acids that vary in their molecular weight and dissociation constants. Due to the relatively high concentration of these ampholytes, IEF electrophoresis requires high voltage and requisite cooling of the agarose matrix. The anode and cathode of the electrophoretic system are placed in acidic and basic environments, respectively. During electrophoresis, the negatively charged proteins migrate toward the anode and the positively charged proteins migrate to the cathode.

After protein migration, the AAT-specific protein bands are visualized by immunochemical detection. Enzyme-labeled anti-AAT antiserum is applied over the gel and allowed to incubate. After washing and blotting to remove unbound antiserum, a substrate is applied to the gel surface that is catalyzed to a colored product by the enzyme label thereby revealing the AAT protein bands.

Prior to the availability of a commercially available AAT phenotyping test, analysis was performed by laboratory-developed tests that most often involved polyacrylamide IEF electrophoresis with visualization of AAT variants using Coomassie Blue staining. In June 2007, the US Food and Drug Administration (FDA) cleared a commercially available reagent system for AAT phenotyping that utilizes agarose IEF electrophoresis and immunostaining (Hydragel 18 A1AT Isofocusing kit, Sebia USA, Norcross, GA). This method is now used by the majority of clinical labs that offer AAT phenotype testing and will be the focus of the analytical component in this chapter.

2   Materials

2.1   *Reagent Kit*

Hydragel 18 A1AT ISOFOCUSING (Cat. #4356), Sebia USA, Inc. The following materials are supplied in the kit.

1. Agarose gels. Support medium for protein isoelectric focusing and immunodetection. Each gel contains 1% agarose and ampholytes necessary for optimum performance. Gels should be stored horizontally in the original protective packaging and refrigerated at 2–8 °C. Prior to use gels should be equilibrated to room temperature.

2. Ethylene glycol solution. Used to provide effective contact between the gel plastic backing and the temperature control plate of the migration module during the electrophoretic migration. Store at room temperature or refrigerated at 2–8 °C.

3. Anodic solution. Used as an electrolyte for preparing anodic strips for isoelectric focusing. Contains proprietary

components necessary for optimum performance. Store at room temperature or refrigerated at 2–8 °C. Solution must be free of precipitate.

4. Cathodic solution. Used as an electrolyte for preparing cathodic strips for isoelectric focusing. Contains sodium hydroxide. Store at room temperature or refrigerated at 2–8 °C. Solution must be free of precipitate. After each use, close the cathodic solution vial immediately and tightly to avoid carbonation of this solution.

5. Sponge strips. Used to ensure contact between the gel and electrodes and determine the pH range during focalization. Moist sponges in the original protective packaging can be stored at room temperature or refrigerated at 2–8 °C but must not be frozen.

6. Sample diluent. Contains saline solution supplemented with bovine serum albumin and sodium azide. Store refrigerated at 2–8 °C. Diluent must be free of precipitate.

7. Antiserum diluent. For diluting peroxidase-labeled anti-AAT antiserum just before use. Contains polyethylene glycol octyl-phenol ether. Store at room temperature or refrigerated at 2–8 °C. Antiserum diluent must be free of precipitate.

8. Wash solution. For washing the agarose gel after the blotting step that follows the incubation with the peroxidase-labeled anti-ATT antiserum. Contains proprietary buffer solution at pH 10.5 ± 0.5 necessary for optimum performance. Store at room temperature or refrigerated at 2–8 °C. Wash solution must be free of precipitate.

9. Rehydrating solution. For rehydrating the agarose gel before the peroxidase-immunostaining step. Contains proprietary components necessary for optimum performance. Store at room temperature or refrigerated at 2–8 °C. Solution must be free of precipitate.

10. TTF1/TTF2 solvent. For the preparation of TTF visualization solution. Contains ethylene glycol and other proprietary components necessary for optimum performance. Store at room temperature or refrigerated at 2–8 °C. TTF1/TTF2 solvent must be free of precipitate.

11. TTF1 stock solution. For visualization of the phenotypes of AAT proteins separated by IEF electrophoresis and labeled via the peroxidase-labeled antiserum. Contains dimethyl sulfoxide and fluorene-2,7-diyldiamine. Store refrigerated at 2–8 °C. TTF1 solution must be free of precipitate.

12. TTF2 stock solution. For visualization of the phenotypes of AAT proteins separated by IEF electrophoresis and labeled via the peroxidase-labeled antiserum. Contains dimethyl

sulfoxide. Store refrigerated at 2–8 °C. TTF1 solution must be free of precipitate.

13. Applicators. Precut, single-use applicators for specimen application onto gel. Store in a dry place at room temperature or refrigerated at 2–8 °C.

14. Thin filter paper. Precut, single-use, thin absorbent paper pads for blotting excessive moisture off the gel surface before specimen application. Store in a dry place at room temperature or refrigerated at 2–8 °C.

15. Thick filter papers. Single-use, thick absorbent paper pads for blotting unprecipitated proteins off the gel after immunostaining, washing, and rehydration steps, and for the removal of excessive substrate reagent. Store in a dry place at room temperature or refrigerated at 2–8 °C.

2.2 Items Not Supplied in the Kit

1. Antiserum. Peroxidase-labeled anti-AAT antiserum for immunostaining. Reconstitute with exactly 0.55 mL of distilled water and allow to stand for 5 min then mix gently. Stable for 1 week refrigerated at 2–8 °C.

2. Hydrogen peroxide (30%). Store in a dark bottle and refrigerated at 2–8 °C.

3. Destaining solution. For washing the gel after the final gel processing. Dilute each vial of destaining solution up to 100 L with distilled water. Diluted destaining solution is stable for 1 week at room temperature in a closed bottle.

4. HYDRASYS Wash solution. For cleaning the HYDRASYS staining compartment. Dilute each vial up to 5 L with distilled water. The diluted wash solution contains an alkaline buffer of pH 8.5 ± 0.5 and sodium azide. Store in a closed containers at room temperature or refrigerated at 2–8 °C.

5. HYDRASYS FOCUSING System.

6. Dynamic mask for antiserum application.

7. Application mask for reagent application.

3 Methods

The HYDRASYS system is a semi-automated multiparameter instrument. The automated steps include processing of agarose gels in the following sequence: specimen application, IEF electrophoresis, incubation with enzyme-labeled antiserum, incubation with enzyme substrate, blotting, drying, washing, and drying of the gel. The manual steps include handling specimens and gels, application of reagents, and setting up the instrument for operation.

3.1 Specimen Preparation

1. For serum samples with an AAT concentration ≤200 mg/dL, dilute the serum 1:10 dilution using the sample diluent (*see* **Note 7**). For serum samples with an AAT concentration >200 mg/dL, dilute the serum 1:20 dilution using the sample diluent. Quality control samples of known AAT phenotype (e.g., MM, SS, and ZZ) should be included on each gel to facilitate interpretation of patient AAT phenotypes (*see* **Note 8**).

3.2 Migration Setup

1. Switch on HYDRASYS instrument and turn the high voltage switch to high voltage mode.

2. Apply the diluted samples to the wells of an 18-teeth applicator and place it with the teeth up into a wet storage chamber.

3. Leave the applicator in the wet storage chamber at room temperature for 5 min. The specimen proteins will then focus into the tip of the applicator teeth and concentrate by partial evaporation.

4. Open the lid of the migration module, raise the electrode and applicator carriers, then select the "A1AT FOCUS TTF" migration program from the instrument menu.

5. Prepare electrode strips by saturating two sponges with anodic and cathodic solutions, respectively, 5 min before use. These are then attached to the appropriate electrodes on the carriers.

6. Streak 300 μL of ethylene glycol across the lower third of the HYDRASYS temperature control plate in the migration module.

7. Place the agarose gel, plastic side down, on a paper towel to remove excess water then apply one thin filter paper onto the gel surface to absorb excess fluid. Remove the paper immediately.

8. Place the gel with the gel side up with its edge against the stop on the plate ensuring that the ethylene glycol distributes evenly under the gel and that no air bubbles are trapped.

9. Remove the applicator from the wet storage chamber and place it in the migration module.

10. Close the lid of the migration module and press the green arrow start button to start the migration procedure.

3.3 Immunostaining

1. After migration, raise the lid of the migration module and remove the applicator and sponges but leave the gel in place.

2. Prepare a working solution of anti-AAT antiserum by adding 40 μL of antiserum to 300 μL of antiserum diluent and add it to the dynamic mask.

3. Apply the antiserum to the surface of the gel by moving the dynamic mask up and down the entire length of the gel.

4. Leave the dynamic mask on the gel in the migration module, close the lid of the migration module, and press the green arrow start button to begin the 10 min incubation phase.

3.4 Gel Blotting

1. After incubation, raise the lid of the migration module and remove the dynamic mask.

2. Apply a thick filter paper, smooth side down, to the surface of the gel and press firmly to ensure adherence.

3. Close the lid of the migration module and press the green arrow start button to begin the 3 min blotting phase.

3.5 Gel Washing

1. After blotting, raise the lid of the migration module and remove the filter paper from the gel surface but leave the gel in place.

2. Place the application mask over the gel.

3. Pipet 7 mL of wash solution through the hole in the application mask and ensure that the entire gel surface is covered by the wash solution.

4. Close the lid of the migration module and press the green arrow start button to begin the 5 min washing phase.

5. After 5 min open the lid of the migration module and remove the wash solution by pipetting it out from the application mask.

6. Remove the application mask but leave the gel in place.

7. Blot the gel as described in Subheading 3.4.

3.6 Gel Rehydration

1. After blotting, raise the lid of the migration module and remove the filter paper from the gel surface but leave the gel in place.

2. Place the application mask over the gel.

3. Pipet 7 mL of rehydrating solution through the hole in the application mask and ensure that the entire gel surface is covered by the rehydrating solution.

4. Close the lid of the migration module and press the green arrow start button to begin the 5 min rehydrating phase.

5. After 5 min open the lid of the migration module and remove the rehydrating solution by pipetting it out from the application mask.

6. Remove the application mask but leave the gel in place.

7. Blot the gel as described in Subheading 3.4.

8. Repeat the rehydration step a second time as described in this section but do not remove the application mask at the end.

3.7 Visualization

1. Prepare the visualization solution by combining together the following, in order: 4 mL TTF1/TTF2 Solvent, 100 μL TTF1, 100 μL TTF2, and 4 μL hydrogen peroxide (30%).

2. Pipet 3.5 mL of visualization solution through the hole in the application mask and ensure that the entire gel surface is covered by the visualization solution.

3. Close the lid of the migration module and press the green arrow start button to begin the 10 min visualization phase.

4. After 10 min open the lid of the migration module and remove the visualization solution by pipetting it out from the application mask.

5. Remove the application mask and blot the gel as described in Subheading 3.4.

3.8 Gel Drying

1. After blotting, raise the lid of the migration module and remove the filter paper from the gel surface, but leave the gel in place.

2. Close the lid of the migration module and press the green arrow start button to begin the 3 min drying phase.

3.9 Wash and Final Processing of the Gel

After drying, the gel must be washed in the integrated staining compartment of the HYDRASYS.

1. Remove the dried gel from the migration module, place it into the gel holder, and place the gel holder into the staining compartment.

2. Add 400 mL of destaining solution and 4600 mL distilled water to the destaining solution container.

3. Select the "WASH ISOENZ/GEL" program from the instrument menu and press the green arrow start button to begin the final processing of the gel.

4. When completed, the gel is ready for evaluation.

4   Notes

The analytical methods for AAT phenotyping are relatively simple because the reagents and apparatus utilized are FDA-cleared for diagnostic testing. However, there is appreciable complexity to interpreting the AAT phenotype from the resulting banding. To facilitate proper identification of various AAT phenotypes this section focuses on interpretation.

4.1 Nomenclature

1. AAT variants are generally named with the prefix Pi to indicate protease inhibitor.

2. Pi is followed with a capital roman letter, which indicates the variant's IEF migration pattern relative to the PiM (migrates in the middle of the gel), PiF (migrates fast or toward the anodal side of the gel), PiS (migrates slow or toward the cathodal side of the gel), and PiZ (migrates very slow or "last").

3. Some AAT variants are coded by a single allele, while others are encoded by multiple alleles. The wild-type variant, PiM, can be

coded by at least three different alleles, which are phenotypically referred to as PiM_1, PiM_2, and PiM_3.

4. Each variant produces multiple bands associated with its visualized phenotype. These bands are defined as either major or minor bands depending on their average staining intensity (major bands generally stain darker). The bands are numbered 2, 4, 6, 7, and 8 from anode to cathode. The minor bands are defined as bands 2, 7, and 8; the major bands are defined as 4 and 6. Not all variants will resolve into 5 bands using IEF.

5. The numeric coding for the different M alleles should not be confused with the numeric coding for the multiple major and minor bands associated with each variant.

6. The term PiM_{null} is often used to indicate a variant that has only wild-type variant visible on the IEF gel, but has a total AAT concentration significantly less than the lower limit of the appropriate reference interval. Exact AAT concentrations associated with the PiM_{null} phenotype have not been defined.

4.2 Interpretation of a Reference Gel (Fig. 1)

1. AAT phenotype determination requires both the sample AAT pattern as resolved by IEF electrophoresis and the total AAT concentration, which is typically quantified using immunoturbidimetric or nephelometric techniques. The total AAT concentrations associated with the reference gel (Fig. 1) are listed in Table 1.

2. All gels should include migration control samples confirmed to contain PiM, PiS, and PiZ variants. These controls act as migration markers to compare all unknown patient samples.

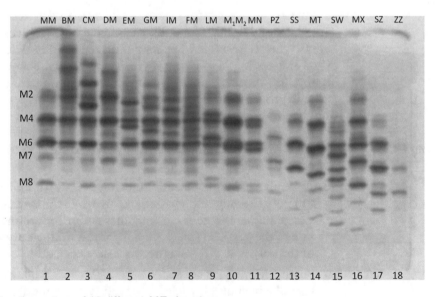

Fig. 1 Migration pattern of 18 different AAT phenotypes

Table 1
Total AAT concentrations associated with the samples visualized in Fig. 1. The established reference range for the immunoturbidimetric assay used is 90–200 mg/dL

Lane	AAT phenotype	AAT concentration (mg/dL)
1	MM	126
2	BM	156
3	CM	209
4	DM	188
5	EM	112
6	GM	208
7	IM	163
8	FM	186
9	LM	156
10	M_1M_2	168
11	MN	82
12	PZ	66
13	SS	116
14	MT	124
15	SW	173
16	MX	156
17	SZ	54
18	ZZ	22

3. It is useful to save any samples confirmed to contain additional common and/or rare variants. These samples can also be used as controls and are particularly useful when rare or new variants are encountered.

4. The wild-type variant, PiMM is shown in lane 1.There is distinct resolution of five bands, which are referred to consecutively as M2, M4, M6, M7, and M8 with M2 being the most anodal band and M8 the most cathodal.

5. The most anodal migrating variant is called PiB (lane 2). The B2 minor band migrates to the anodal most tip of the gel. The B4 and B6 major bands both migrate anodal to the M2 minor band. The B7 and B8 minor bands migrate identical to the C4 and C6 major bands (lane 3).These comigrations can cause some confusion, but the staining intensity of the bands should be distinct for the two phenotypes.

6. The PiC variant can be identified by the C4 and C6 major bands that flank the anodal and cathodal sides of the M2 band (lane 3). The PiC phenotype also has a C2 minor band that usually stains with dark intensity anodal to the C4 band.

7. The PiD variant migrates very similarly to PiM with a few distinct characteristics (lane 4).The D2 and M2 minor bands resolve, with the D2 band migrating anodal to the M2 band. Less resolved are D4/M2, D6/M4, and D8/M7. The D7 is resolved, migrating between M4 and M6.

8. Care should be taken not to mistake the PiE phenotype for a PiS phenotype (lane 5 versus lane 13).The PiE minor bands (E2, E7, and E8) migrate to the same location as the PiS major bands (S4 and S6). PiE can be distinguished from PiS because the major bands E4 and E6 migrate slightly cathodal to M2 and M4, respectively.

9. PiG and PiI migrate identically (lanes 6 and 7) and cannot be reliably distinguished from each other without a total AAT concentration and/or DNA sequencing. PiI is coded from a deficiency allele and therefore if the patient sample has an AAT concentration less than the reference interval the variant is likely PiI. However, if the concentration is within the reference interval PiG cannot be assumed because the concentration might be transiently elevated due to an acute phase response. All but one of the major and minor bands for the PiG and PiI variants migrate slightly cathodal to the PiM bands. The G2 and I2 bands migrate anodal to the M2 band. An additional subtle difference between the PiI and PiG variants is in the intensity of band staining. In the G phenotype the G4 and G6 bands stain with equal intensity. In contrast, the I4 band stains darker than the I6 band.

10. The PiF variant (lane 8) can be distinguished from the others by its distinct pair of doublets found just cathodal and anodal to the M4 band. The PiF variant is a functional deficiency, meaning that the AAT concentration in the sample should fall within the reference interval (Table 1), but the clinical phenotype is often pathological.

11. PiL has a similar migration pattern to PiP (lane 9 compared to 12), but has an additional major band (L4) visible just anodal to M4.

12. The wild-type PiM variant can be coded by at least three different alleles and lead to the M subtypes M_1, M_2, and M_3, which are clinically equivalent. The wild-type subvariants have identically migrating major bands, but slight differences in the minor bands. The M_3 minor bands migrate directly between the M_1 and M_2 minor bands. Phenotypically distinguishing between these variants is not necessary, but it is important to

recognize that they occur so that there is no confusion when PiM specimens migrate slightly different.

13. The PiN variant is not a deficiency variant, but the N7 and N8 minor bands migrate virtually identical to the Z4 and Z6 major bands (lane 11 compared to 12). The major N bands are easy to recognize because they make a distinct doublet pattern with the M4 and the M6 bands, with the N4 and N6 bands migrating more slowly leaving them closer to the cathode.

14. PiP has one major band centered between M4 and M6 and a second major band that comigrates with M7 (lane 12). As mentioned previously, PiP can be distinguished from PiL by the absence of a major band anodal to M4.

15. The decreased serum concentration of the PiZ variant makes it stain weaker than most variants. The M4 and M6 major bands stain with similar intensity to the M minor bands and migrate just cathodal to M7 and M8 (lanes 12, 17, and 18).

16. Although PiS is the second most common deficiency allele, PiS will often still stain with dark intensity, and in homozygotes all five bands are visible (lane 13). The S4 band migrates just slightly anodal to the M6, while the S6 and S7 bands migrate between M7 and M8. The S8 migrates slowly, and remains close to the cathode, terminating migration before the Z6 band (lanes 13 and 17). Identification of PiT will likely require the specimen in question to be run in a lane juxtaposing a PiS control sample. This is because all of the major and minor T bands migrate directly cathodal to the S bands (lane 14).

17. The PiW variant can be identified by locating the W4 and W6 major bands that migrate anodal to the Z4 and Z6 major bands, but cathodal to the S4 and S6 major bands (Lane 15).

18. The PiX variant migrates similarly to the PiW variant, but following the conventional nomenclature the corresponding bands migrate slower and remain slightly more cathodal.

4.3 Miscellaneous Notes

1. Additional descriptions of variants not visualized in this gel can be found in the literature [2–5].

2. Unknown variants are sometimes encountered when analyzing specimens using IEF electrophoresis. In these cases, gene sequencing can be used to identify the underlying mutation(s) as clinically indicated.

3. Total AAT concentrations associated with the various phenotypes observed in the gel are outlined in Table 1. Population studies have been published to evaluate the AAT concentrations expected for the many common and rare phenotypes [6, 7].

References

1. Burtis C, Ashwood E, Bruns D (eds) (2011) Tietz textbook of clinical chemistry and molecular diagnostics, 5th edn. Elsevier, Amsterdam

2. Zerimech F, Hennache G, Bellon F et al (2008) Evaluation of a new Sebia isoelectrofocusing kit for alpha 1-antitrypsin phenotyping with the Hydrasys system. Clin Chem Lab Med 46:260–263

3. Greene D, Elliott-Jelf M, Straseski J et al (2013) Facilitating the laboratory diagnosis of α1-antitrypsin deficiency. Am J Clin Pathol 139:184–191

4. Greene D, Procter M, Krautscheid P et al (2012) α1-antitrypsin deficiency in fraternal twins born with familial spontaneous pneumothorax. Chest 141:239–241

5. BB S-L, Procter M, Krautscheid P et al (2014) Challenging identification of a novel PiISF and the rare PiMmaltonZ α1-antitrypsin deficiency variants in two patients. Am J Clin Pathol 141:742–746

6. Bornhorst J, Greene D, Ashwood E et al (2013) α1-antitrypsin phenotypes and associated serum protein concentrations in a large clinical population. Chest 143:1000–1008

7. Donato L, Jenkins S, Smith C et al (2012) Reference and interpretive ranges for α(1)-antitrypsin quantitation by phenotype in adult and pediatric populations. Am J Clin Pathol 138:398–405

Chapter 5

Laboratory Diagnosis by Genotyping

Irene Belmonte, Luciana Montoto, and Francisco Rodríguez-Frías

Abstract

Alpha-1 antitrypsin (AAT) genotyping is useful to confirm the clinical diagnosis of AAT deficiency and determine the specific allelic variant. Genotyping is the reference standard procedure for identifying rare allelic variants and characterizing new variants. It is also useful when there is a discrepancy between the patients' AAT levels and their phenotypes. AAT genotype is determined by an allele-specific genotyping assay for the S, Z, and Mmalton variants and by exome sequencing.

Key words Alpha-1 antitrypsin laboratory diagnosis, Alpha-1 antitrypsin genotyping, Alpha-1 antitrypsin deficiency, Alpha-1 antitrypsin allele-specific genotyping, Alpha-1 antitrypsin exome sequencing, LightCycler, Melting curves

1 Introduction

The current approach to laboratory diagnosis of alpha-1 antitrypsin (AAT) deficiency involves a combination of quantitative AAT measurement by nephelometry and phenotype characterization by isoelectric focusing (IEF) [1]. However, most rare variants are difficult or even impossible to detect with these methods, and this may contribute to misclassification of many of these cases, with subsequent underestimation of their true frequency [2]. Most of these alleles can only be detected by molecular analysis of the AAT gene, using methods such as allele-specific genotyping or exome sequencing [3].

In allele-specific genotyping, a fragment of the AAT gene is amplified by real-time PCR with specific oligonucleotide primers. Fluorescent labeled probes are used to identify the PCR product and determine the genotype by melting curve analysis (FRET technology) [4]. Exome sequencing, which requires complete study of the DNA sequences of the 4 encoding exons (II, III, IV, V) of the AAT gene [5], consists of amplification of the 4 exons by polymerase chain reaction (PCR) followed by cycle sequencing of the PCR products.

Florie Borel and Christian Mueller (eds.), *Alpha-1 Antitrypsin Deficiency: Methods and Protocols*, Methods in Molecular Biology, vol. 1639, DOI 10.1007/978-1-4939-7163-3_5, © Springer Science+Business Media LLC 2017

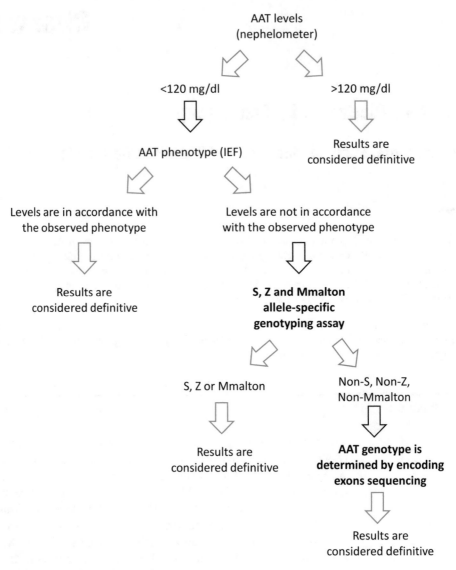

Fig. 1 Proposed AATD diagnostic algorithm

The diagnostic algorithm for AAT deficiency we have developed is performed as follows (Fig. 1): AAT serum levels are measured, and in cases with concentrations lower than 120 mg/dL, the phenotype is determined by IEF. If AAT values concur with the phenotype observed, the laboratory results are considered definitive [6]. However, if AAT levels do not correspond to the phenotype, the AAT genotype is determined by an allele-specific genotyping assay for the S, Z, and Mmalton variants and by exome sequencing. Allele-specific determination of the S and Z variants is done with a commercial kit (Tib Molbiol, Berlin, Germany), but Mmalton detection is performed using primers and probes designed in the laboratory [7, 8].

Molecular analysis of the AAT gene or genotype is useful in cases of discrepancy between the patients' AAT levels and their phenotypes. Genotyping is the reference standard for identifying rare allelic variants associated with hereditary AAT deficiency and for characterizing new variants [5]. It is also the most appropriate assay for identifying null variants [9].

Lately, we have been focusing a great deal of attention on the issue of AAT genotyping, and we have developed several strategies to improve molecular analysis of the AAT gene using dried blood spot (DBS) and serum samples. Here, we describe the methods we now use for AAT genotyping in whole blood, DBS, and serum samples.

2 Materials

2.1 DNA Extraction

1. QIAamp DNA Mini Kit (Qiagen, Hilden, Germany).
2. Sterile water.
3. Puncher

2.2 Allele-Specific Genotyping Assay

1. LightCycler 2.0 Instrument (Roche Diagnostic, Mannheim, Germany).
2. For LightCycler 2.0: LightCycler capillaries (20 μL), LightCycler sample carousel, LightCycler centrifuge adapters, and LightCycler capping tool.
3. LightMix in vitro diagnostics kit, Alpha-1 Antitrypsin (AAT) Pi*S and Pi*Z (Tib Molbiol, Berlin, Germany).
4. LightCycler FastStart DNA Master Hybprobe (Roche Diagnostic, Mannheim, Germany): LightCycler FastStart Enzyme, LightCycler FastStart Reaction Mix HybProbe, $MgCl_2$ Stock Solution 25 mM, and water.
5. Primers and probes for Mmalton detection (GenBank accession no. K02212) (Table 1).

Table 1
Primer and probe sequences and nucleotide positions for Mmalton allele-specific genotyping

Primer and Probe	Position	Sequence
Forward primer	7454–7475	5′-TTCAACAAGATCACCCCC AA CC-3′
Reverse primer	7604–7624	5′-GCCCTCCAGGATTTCATC GTG-3′
Anchor probe	7506–7528	5′-AGCTGGCACACCAGTCCAACAGC-FL-3′
Sensor probe (specific for mutated sequence)	7530–7554	5′-LC640-CCAATATCTTCTCCCCAGTGAG-PH-3′

Table 2
Primer sequences, nucleotide positions, sizes, and annealing temperatures for PCR and exome sequencing of AAT

Exon	Sequence	Position	Size	Annealing temperature
2	IIAFw: 5′ GATCACTGGGAGTCATCATGTGC 3′	7251–7273	450	54
	IIARv: 5′ GGTTGAGGGTACGGAGGAGT 3′	7681–7700		
	IIBFw: 5′CCAAGGCTGACACTCACGAT 3′	7590–7609	455	
	IIBRv: 5′AGGAGAGTTCAAGAACTGATG GTT 3′	8021–8044		
3	IIIFw: 5′ TTCCAAACCTTCACTCACCCCT GGT 3′	9362–9386	552	60
	IIIRv: 5′ CGAGACCTTTACCTCCTCACCC TGG 3′	9889–9913		
4	IVFw: 5′ CCCAGAAGAACAAGAGGAATGC TGT 3′	10,882–0906	257	54
	IVRv: 5′ CATTCTTCCCTACAGATACCA TGGT 3′	11,114–11,138		
5	VFw: 5′ TGTCCACGTGAGCCTTGCTCGA GGC 3′	11,841–11,865	339	54
	VRv: 5′ GACCAGCTCAACCCTTCTTTAA TGT 3′	12,155–12,179		

2.3 Exome Sequencing

1. FastStart High Fidelity PCR System, dNTPack (Roche Diagnostic, Mannheim, Germany): FastStart High Fidelity Enzyme Blend (5 U/μL) in storage buffer, FastStart High Fidelity Reaction Buffer concentrated 10× with 18 mM MgCl$_2$, and PCR Nucleotide Mix.

2. Primers for PCR and sequencing of the AAT exome (GenBank accession no. K02212) (Table 2):

3. 0.2 mL 8-strip tubes and caps.

4. ExoSAP-IT for PCR Product Cleanup (Affymetrix, Santa Clara, CA).

5. BigDye Terminator v3.1 Cycle Sequencing Kit (Life Technologies, Madrid, Spain): Big Dye Terminator v3.1 Ready Reaction Mix, and 5× Sequencing Buffer.

6. BigDye XTerminator Purification Kit (Life Technologies, Madrid, Spain).

7. Sequencing plates.

8. Plate septa 96 well (Life Technologies, Madrid, Spain).

9. 3130/3130xl Genetic Analyzer (Applied Biosystems, Foster City, CA).

3 Methods

3.1 DNA Extraction from Whole Blood and Serum Samples

1. Perform DNA extraction from whole blood or serum sample according to QIAamp DNA Mini Kit protocol (*see* **Notes 1** and **2**).

3.2 DNA Extraction from DBS Samples

1. Cut a 3 mm paper DBS disk with a puncher and place it in a 1.5 mL plastic tube (*see* **Note 3**). Add 80 μL of water and vortex. Incubate the mix at 60 °C overnight.

2. Add 50 μL of water and incubate at 80 °C for 30 min. Centrifuge during 10 min at 6000 × g to eliminate contaminants (*see* **Note 4**).

3. Recover the eluted DNA with a pipette in a new 1.5 mL plastic tube (*see* **Note 5**).

3.3 Allele-Specific Genotyping Assay for S and Z Variants

1. In a 1.5 mL reaction tube on ice, prepare the PCR mix according to the LightMix in vitro diagnostics kit Alpha-1 Antitrypsin (AAT) Pi*S and Pi*Z protocol for the LightCycler 2.0 Instrument (*see* **Note 6**). Mix carefully by pipetting up and down (*see* **Note 7**) and store on ice.

2. Place the capillaries into the LightCycler centrifuge adapters and pipet 8 μL of PCR mix into the plastic reservoir at the top of the capillary. Add 2 μL of DNA template to the capillary.

3. Seal each capillary with a plastic stopper using the LightCycler capping tool. With the help of the centrifuge adapters, place the capillaries in a benchtop centrifuge and spin down the PCR mix inside the capillaries.

4. Place the capillaries in the LightCycler sample carousel, keeping them in an upright position (*see* **Note 8**).

5. Place the LightCycler sample carousel in the LightCycler 2.0 Instrument and start the PCR program described in the LightMix in vitro diagnostics kit Alpha-1 Antitrypsin (AAT) Pi*S and Pi*Z protocol.

6. The melting curves and melting peaks are obtained following the instructions in the instrument manual (Figs. 2, 3). The S allele is detected in channel 530. It shows a high melting temperature for the S allele and a lower melting temperature for the wild-type allele. Channel 640 shows the Z allele, which has a high melting temperature for the wild-type allele and a lower melting temperature for the Z allele. The result of this assay must always show 1 or 2 melting peaks for each channel in any of the possible combinations:

   In channel 530 at 50 °C ± 2.5 °C (wild type)/55 °C ± 2.5 °C (S allele).

   In channel 640 at 55 °C ± 2.5 °C (Z allele)/62 °C ± 2.5 °C (wild type).

3.4 Allele-Specific Genotyping Assay for the Mmalton Variant

1. In a 1.5 mL reaction tube on ice, prepare the PCR mix for one 20 μL reaction by adding the following components in the order listed in Tables 1 and 3 (*see* **Note 9**). Mix carefully by pipetting up and down (*see* **Note 7**) and store on ice.

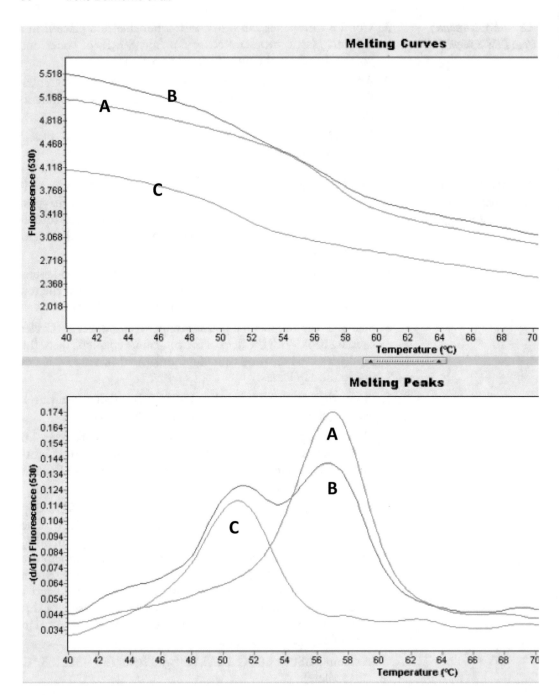

Fig. 2 Melting curves and melting peaks obtained in channel 530, containing the different S genotype profiles: (**a**) homozygous mutated sample, (**b**) heterozygous sample, and (**c**) wild-type sample

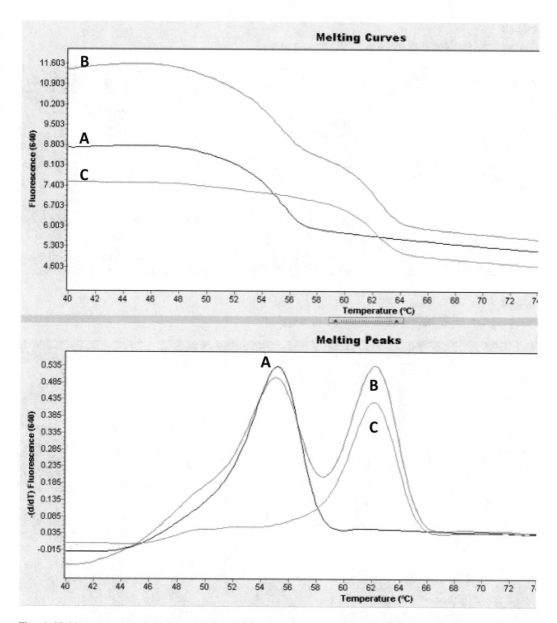

Fig. 3 Melting curves and melting peaks obtained in 640, containing the different Z genotype profiles: (**a**) homozygous mutated sample, (**b**) heterozygous sample, and (**c**) wild-type sample

2. Place the capillaries in the LightCycler centrifuge adapters and pipet 15 μL of PCR mix into the plastic reservoir at the top of the capillary. Add 5 μL of DNA template to the capillary.

3. Seal each capillary with a plastic stopper using the LightCycler capping tool. With the help of the centrifuge adapters, place the capillaries in a benchtop centrifuge and spin down the PCR mix inside the capillaries.

Table 3
Components for Mmalton allele-specific genotyping

Component	Volume (μL)
Water	7.4
MgCl$_2$ (25 mM)	1.6
Forward primer (10 μM)	1
Reverse primer (10 μM)	1
Anchor (4 μM)	1
Sensor (4 μM)	1
LightCycler FastStart Enzyme +60 μL Reaction Mix HybProbe	2
Total volume	15

Table 4
PCR program for Mmalton allele-specific genotyping

Step	1	2			3			4
Analysis mode	None	Quantification mode			Melting Curves mode			None
Cycles	1	45			1			1
Target (°C)	95	95	54	72	95	40	85	40
Acquisition mode	None	None	Single	None	None	None	Cont.	None
Hold (hh:mm:ss)	00:10:00	00:00:05	00:00:10	00:00:15	00:00:20	00:00:20	00:00:00	00:00:30
Ramp rate (°C/s)	20	20	20	20	20	20	0.2	20
Sec target (°C)	0	0	0	0	0	0	0	0
Step size (°C)	0	0	0	0	0	0	0	0
Step delay (cycles)	0	0	0	0	0	0	0	0

4. Place the capillaries in the LightCycler sample carousel, keeping them in an upright position (*see* **Note 8**).

5. Place the LightCycler sample carousel in the LightCycler 2.0 instrument and start the PCR program described below (Table 4).

6. The melting curves and melting peaks are obtained following the instructions in the instrument manual (Fig. 4). In channel 640, the Mmalton allele is detected, with a high melting temperature for the Mmalton allele and a lower melting

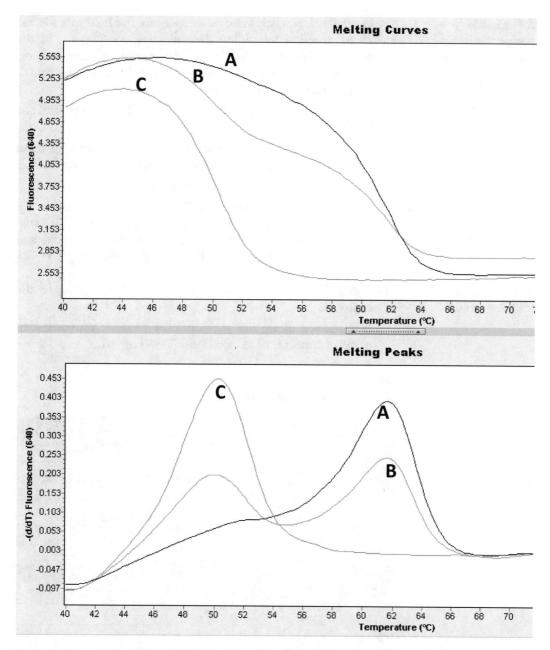

Fig. 4 Melting curves and melting peaks containing the different Mmalton genotype profiles: (**a**) homozygous mutated sample, (**b**) heterozygous sample, and (**c**) wild-type sample

temperature for the wild-type allele. The result of this assay must always show one or two melting peaks at 50 °C ± 2.5 °C/62 °C ± 2.5 °C. The peak at 62 °C indicates that the Mmalton allele is present and the peak at 50 °C indicates that the allele is absent.

3.5 Exome Sequencing Using Whole Blood Samples

3.5.1 Amplification of the 4 Encoding Exons of the AAT Gene

1. In a 1.5 mL reaction tube on ice, prepare the PCR mix for one 25 µL reaction by adding the following components in the order mentioned below (Table 5) (*see* **Note 9**). Prepare one mix for each exon (II, III, IV and V), using a specific pair of primers for each one (*see* Table 2).

2. Mix carefully and pipet 15 µL PCR mix into each well of the 8-strip tubes.

3. Add 10 µL of DNA template. Cap the tubes and spin them down. Transfer the tubes to the thermocycler (*see* **Note 10**).

4. Thermocycling conditions are shown in Table 6:

5. Perform electrophoresis using 2% agarose gel to visualize the PCR product. Load 5–10 µL of DNA (*see* **Note 11**). This step enables determining whether the PCR was successful and whether the resulting product is the correct size (Fig. 5).

3.5.2 Purification of the PCR Product

1. Purify the PCR product using ExoSAP-IT according to the protocol.

2. If necessary, dilute the purified PCR product with water to obtain a concentration of around 30–40 ng/µL.

Table 5
Components for encoding AAT gene amplification

Component	Volume (µL)
Water	9.75
FastStart High Fidelity Reaction Buffer 10×	2.5
PCR Nucleotide Mix (10 Mm each)	0.5
Forward primer (10 µM)	1
Reverse primer (10 µM)	1
FastStart High Fidelity Enzyme Blend (5 U/µL)	0.25
Total volume	15

Table 6
PCR program for encoding AAT gene amplification

Step	Temperature (°C)	Time (hh:mm:ss)	Cycles
Denaturation	95	00:02:00	1
Amplification	95	00:00:30	35
	Primers annealing (54/60)	00:00:30	
	72	00:00:30	
Final extension	72	00:05:00	1
Cooling	4	∞	1

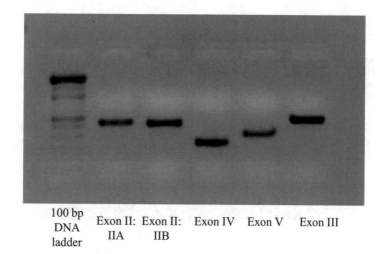

100 bp
DNA   Exon II: Exon II: Exon IV   Exon V   Exon III
ladder   IIA   IIB

Fig. 5 Checking the PCR product on a 2% agarose gel

Table 7
Components for AAT gene sequencing

Component	Volume (µL)
Water	6
5× Sequencing buffer	1.5
Primer (10 µM)	0.5
Big Dye Terminator v3.1 Ready Reaction Mix	1
Total volume	9

3.5.3 AAT Sequencing

1. In a 1.5 mL reaction tube on ice, prepare the PCR mix for one 10 µL reaction by adding the following components in the order listed below (Table 7) (*see* **Notes 9** and **12**). Prepare one mix for each exon (II, III, IV and V) and for each sequence direction (Forward and Reverse) using the different primers (*see* Table 2). There will be ten mixes in total.

2. Mix carefully and pipet 9 µL of mix into each well of the 8-strip tubes.

3. Add 1 µL of the purified PCR product and close the strip tubes with the caps. Spin down the tubes and transfer them to the thermocycler (*see* **Note 10**).

4. Start the PCR program described in Table 8.

5. Transfer the sequencing product to the sequencing plates.

6. Purify the sequencing products using the BigDye XTerminator Purification Kit according to the protocol.

Table 8
PCR program for AAT gene sequencing

Step	Temperature (°C)	Time (hh:mm:ss)	Cycles
Amplification	96 Primers annealing (54/60) 60	00:00:10 00:00:05 00:04:00	25
Cooling	4	∞	1

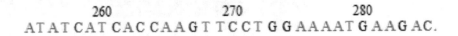
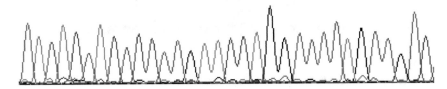

Fig. 6 Sequence fragment obtained in AAT exome sequencing

7. Seal the sequencing plate with plate septa.

8. Spin down at $600 \times g$.

9. Perform sequence analysis of the AAT exons in the 3130xl Genetic Analyzer (Fig. 6) (*see* **Note 13**).

3.6 Exome Sequencing Using DBS Samples

3.6.1 Amplification of the 4 Encoding Exons of AAT

1. In a 1.5 mL reaction tube on ice, prepare the PCR mix for one 25 μL reaction by adding the following components in the order listed below (Table 9) (*see* **Note 9**). Prepare one mix for each exon (II, III, IV and V) using a specific pair of primers for each one (*see* Table 2).

2. Mix carefully and pipet 10 μL of PCR mix into each well of the 8-strip tubes.

3. Add 15 μL of DNA template and cap the strip tubes. Spin down the tubes and transfer them to the thermocycler (*see* **Note 10**).

4. Thermocycling conditions are the same as those used in AAT amplification with DNA from whole blood samples (*see* Table 6).

5. Perform electrophoresis with 2% agarose gel to visualize the PCR product. Load 5–10 μL of DNA (*see* **Note 11**).

3.6.2 Purification of the PCR Product

1. Repeat step 3.5.2

3.6.3 AAT Sequencing

1. Repeat step 3.5.3.

Table 9
Components for encoding AAT gene amplification

Component	Volume (µL)
Water	4.75
FastStart High Fidelity Reaction Buffer 10×	2.5
PCR Nucleotide Mix (10 Mm each)	0.5
Forward primer (10 µM)	1
Reverse primer (10 µM)	1
FastStart High Fidelity Enzyme Blend (5 U/µL)	0.25
Total volume	10

Table 10
Components for first round of PCR

Component	Volume (µL)
Water	2.37
FastStart High Fidelity Reaction Buffer 10×	1.25
PCR Nucleotide Mix (10 Mm each)	0.25
Forward primer (10 µM)	0.5
Reverse primer (10 µM)	0.5
FastStart High Fidelity Enzyme Blend (5 U/µL)	0.13
Total volume	5

3.7 Exome Sequencing Using Serum Samples

3.7.1 Amplification of the 4 Encoding Exons of AAT

1. In a 1.5 mL reaction tube on ice, prepare PCR mix for one 12.5 µL reaction by adding the following components in the order listed in Table 10 (*see* **Note 9**).

2. Mix carefully and pipet 5 µL PCR mix into each well of the 8-strip tubes.

3. Add 7.5 µL of DNA template and cap the tubes. Spin down the tubes and transfer them to the thermocycler (*see* **Note 10**).

4. Thermocycling conditions are the same as those used in AAT amplification with DNA from whole blood samples (*see* Table 6).

5. For a second round of PCR (*see* **Note 14**), prepare the 20 µL PCR mix in a 1.5 mL reaction tube on ice by adding the components described below (Table 11):

6. Mix carefully and pipet 19 µL PCR mix into each well of the 8-strip tubes.

Table 11
Components for second round of PCR

Component	Volume (μL)
Water	14.8
FastStart High Fidelity Reaction Buffer 10×	2
PCR Nucleotide Mix (10 Mm each)	0.4
Forward primer (10 μM)	0.8
Reverse primer (10 μM)	0.8
FastStart High Fidelity Enzyme Blend (5 U/μL)	0.2
Total volume	19

Table 12
PCR program for second round of PCR

Step	Temperature (°C)	Time (hh:mm:ss)	Cycles
Denaturation	95	00:02:00	1
Amplification	95 Primers annealing (54/60) 72	00:00:30 00:00:30 00:00:40	20
Final extension	72	00:05:00	1
Cooling	4	∞	1

7. Add 1 μL of the PCR round 1 product and cap the tubes. Spin down the tubes and transfer them to the thermocycler (*see* **Note 10**).

8. Start the PCR program described in Table 12:

9. Perform electrophoresis with 2% agarose gel to visualize the PCR product. Load 5–10 μL of DNA (*see* **Note 11**).

3.7.2 Purification of the PCR Product

1. Repeat step 3.5.2.

3.7.3 AAT Sequencing

1. Repeat step 3.5.3.

4 Notes

1. In the last step of the DNA extraction protocol, DNA from the whole blood sample is eluted in 80 μL of sterile water and DNA extraction from the serum sample is eluted in 50 μL of sterile water. A fresh or frozen blood sample can be used.

2. The quality of the serum is very important in this step. If it is possible, Perform DNA extraction on the same day the sample is obtained. Do not freeze the serum sample. In addition, it is better to use the DNA on the same day as DNA extraction.

3. The puncher used for this procedure is sterilized each time by flaming to avoid cross contamination of specimens.

4. The quality of the DBS sample is important. Fresh blood should be collected by venotip or fingertip puncture and the amount should completely fill the DBS paper circle. The filter paper should be allowed to air-dry for at least 12 h and after this period, it can be shipped or stored at room temperature for up 7 days or stored at −20 °C for longer periods.

5. Avoid touching the filter paper with the pipette tip.

6. The LightCycler 2.0 instrument allows simultaneous analysis of 32 samples. LightCycler 480 is needed to analyze a larger number of samples (96 samples simultaneously). LightCycler 480 uses plates instead of capillaries.

7. Do not vortex.

8. Make sure that all capillaries are fixed in the optimal position, where the O-ring of the LightCycler sample carousel covers the lower part of the plastic chamber.

9. To prepare the PCR mix for more than one reaction, multiply the amount in the "Volume" column by z, where z = the number of reactions to be run + two additional reactions.

10. Primers for AAT amplification and sequencing have two different annealing temperatures: 54 °C and 60 °C. For AAT amplification or sequencing, a thermocycler that allows simultaneous use of two different annealing temperatures is needed or two different thermocyclers can be used.

11. Do not load the entire volume of PCR product because 5 μL are needed in the next step.

12. Avoid prolonged exposure to light because the reaction mix contains fluorochrome-labeled dideoxynucleotides (ddNTPs) and they can be degraded.

13. Sequence analysis of the AAT exons is applicable to any sequence analyzer.

14. The first and second PCR rounds should be performed on the same day. Do not freeze the first PCR product.

Acknowledgments

This study was supported in part by a grant from Fundación Catalana de Pneumología (FUCAP 2015) and by funding from

Grifols to the Catalan Center for Research in Alpha-1 antitrypsin deficiency of the Vall d'Hebron Research Institute in Vall d'Hebron University Hospital, Barcelona, Spain.

References

1. Molina J, Flor X, García R, Timiraos R, Tirado-Conde G, Miravitlles M (2011) The IDDEA project: a strategy for the detection of alpha-1 antitrypsin deficiency in COPD patients in the primary care setting. Ther Adv Respir Dis 5:237–243

2. Rodriguez-Frias F, Miravitlles M, Vidal R, Camos S, Jardi R (2012) Rare alpha-1-antitrypsin variants: are they really so rare? Ther Adv Respir Dis 6:79–85

3. Costa X, Jardi R, Rodriguez F, Miravitlles M, Cotrina M, Gonzalez C et al (2000) Simple method for alpha1-antitrypsin deficiency screening by use of dried blood spot specimens. Eur Respir J 15:1111–1115

4. Rodriguez F, Jardí R, Costa X, Cotrina M, Galimany R, Vidal R et al (2002) Rapid screening for alpha1-antitrypsin deficiency in patients with chronic obstructive pulmonary disease using dried blood specimens. Am J Respir Crit Care Med 166:814–817

5. Vidal R, Blanco I, Casas F, Jardí R, Miravitlles M (2006) Guidelines for the diagnosis and management of alpha-1 antitrypsin deficiency. Arch Bronconeumol 42:645–659

6. Miravitlles M, Herr C, Ferrarotti I, Jardi R, Rodriguez-Frias F, Luisetti M et al (2010) Laboratory testing of individuals with severe alpha1-antitrypsin deficiency in three European centres. Eur Respir J 35:960–968

7. Belmonte I, Montoto L, Miravitlles M et al (2016) Rapid detection of Mmalton α1-antitrypsin deficiency allele by real-time PCR and melting curves in whole blood, serum and dried blood spot samples. Clin Chem Lab Med 54:241–248

8. Belmonte I, Barrecheguren M, López-Martínez RM, Esquinas C, Rodríguez E, Miravitlles M, Rodríguez-Frias F (2016) Application of a diagnostic algorithm for the rare deficient variant mmalton of alpha-1-antitrypsin deficiency: a new approach. Int J Chron Obstruct Pulmon Dis 11:2535–2541

9. Ferrarotti I, Scabini R, Campo I, Ottaviani S, Zorzetto M, Gorrini M, Luisetti M (2007) Laboratory diagnosis of alpha1-antitrypsin deficiency. Transl Res 150(5):267–274

Chapter 6

Genotyping Protocol for the Alpha-1 Antitrypsin (PiZ) Mouse Model

Alisha M. Gruntman

Abstract

The most common alpha-1 antitrypsin (AAT) mutant variant is a missense mutation (E342K), commonly referred to as PiZ. A transgenic mouse model exists that expresses the mutant human PiZ AAT gene. This protocol outlines the procedure used to extract DNA from and genotype AAT transgenic (PiZ) mice.

Key words Alpha-1 antitrypsin, PiZ, Genotyping, Murine, PCR, DNA extraction

1   Introduction

The most common alpha-1 antitrypsin (AAT) mutant variant is termed PiZ. This mutation leads to decreased serum levels of AAT due to accumulation of a mutant protein in the hepatocyte endoplasmic reticulum. This accumulation can result in liver disease in affected patients. In 1989, a PiZ mouse model was created by inserting the human PiZ AAT gene into the germ line of mice [1]. This insertion resulted in a PiZ dose dependent liver pathology within these mice. However, because the wild-type murine AAT was not knocked out, this mouse model does not recapitulate the human lung disease seen in PiZ AAT patients. In this chapter we describe the genotyping protocol currently used in the PiZ mouse model in our lab. Briefly, we describe the DNA extraction protocol from mouse tissue, performing PCR to amplify the bands of interest, and the running and interpretation of the electrophoresis gel.

2   Materials

2.1 Equipment   Heat block.

PCR thermocycler.

Florie Borel and Christian Mueller (eds.), *Alpha-1 Antitrypsin Deficiency: Methods and Protocols*, Methods in Molecular Biology, vol. 1639, DOI 10.1007/978-1-4939-7163-3_6, © Springer Science+Business Media LLC 2017

Gel box and power source.

Gel camera or viewer.

2.2 Consumables   1.5 ml microcentrifuge tubes.

PCR tube strip with cap strip.

Or

96-well PCR plate.

Thermal seal films for PCR.

2.3 Chemicals/   Alkaline Lysis Reagent (0.2 mM sodium EDTA, 25 mM sodium
Reagents   hydroxide, (NaOH)): 20 ml of 0.5M EDTA, 125 µl of 10N
    NaOH, 50 ml of ddH$_2$O.

Neutralization Buffer (40 mM Tris–HCl): 325 mg of Tris–HCl,
    50 ml of ddH$_2$O.

MasterMix (5 Prime, #2200110).

Molecular grade water.

Genotyping primers.

824: 5′: TTG AGG AGC GAG AGG CAG TT

649: 5′: GAG GCG CTT GTC AGG AAG AT

444: 5′: TGC CAG GAA AGC AAG ATA ACT CTC

645: 5′: TCG AGG AGC ACG ATT CTC TTA TTC

100 bp DNA ladder.

Ethidium lromide.

50× Tri–acetate–EDTA (TAE) Buffer.

Agarose, general purpose LE.

3 Methods

3.1 DNA Extraction   1. Collect mouse tail (0.2 cm) or ear snip/punch (2 mm).
(Based on Truett et al.   2. Place tissue in a 1.5 ml microfuge tube.
[2])
    3. Add 50–100 µl of the Alkaline Lysis Reagent.

    4. Place tubes in a thermomixer set at 95 °C and 500 × g for
       30–60 min.

    5. Cool on ice for 5–10 min.

    6. Spin down samples for 1 min at 8000 × g (this step can be
       excluded if no centrifuge is available, but some of the sample
       may be lost if it has condensed on the lid during incubation).

    7. Add 50–100 µl of Neutralization Buffer and vortex briefly.

    8. There will still be tissue present, but DNA will be in solution.

    9. Store solution at −20 °C until ready to use for PCR.

3.2 Genotyping Procedure

Four primers are needed for this protocol, the first two serve to detect the presence of human AAT (PiZ). The second two primers serve as an internal control and detect a mouse gene (mPMS2). PiZ positive mice should therefore have two detectable bands (*see* Fig. 1 and **Note 1**).

Primer 824 and 649 for PiZ AAT yield a 420 bp band

Primer 444 and 445 for mPMS2 yield a 661 bp band

1. Set up a master mix for the reaction (should be enough for the number of samples plus controls and an additional 10–15% to account for losses during pipetting).

   Amount to add to master mix for each sample to be genotyped (Table 1):

2. Load 2 μl of each of the samples of DNA extracted in Subheading 3.1 above to individual wells of a 96-well PCR plate or individual PCR tubes. Then add 23 μl of the master mix prepared in **step 1** above to each well/tube. Mix by gently pipetting up and down several times.

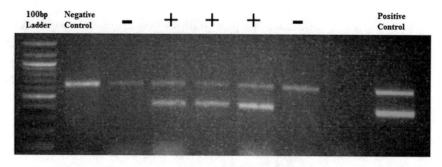

Fig. 1 Example PiZ Mouse DNA genotyping gel. A 100 bp ladder was used as well as a known positive and negative control animal. The − and + symbols indicate PiZ negative and positive animals, respectively

Table 1
Master Mix Preparation

Reagent	Volume
5 Prime Master Mix (2.5×)	10 μl
Primer 824	0.5 μl
Primer 649	0.5 μl
Primer 444	0.5 μl
Primer 445	0.5 μl
Molecular grade water	11 μl
Total per reaction	23 μl

Table 2
Thermocycler Program Settings

Number of cycles	Temperature	Time
1	95 °C	5 min
30	95 °C	30 s
	58 °C	1 min
	72 °C	1 min
1	72 °C	10 min
1	4 °C	Hold

3. Carefully cap the PCR tubes or cover the top of the plate using a seal film (*see* **Note 2**).

4. Run the following program on a thermocycler (*see* **Note 3**) (Table 2).

5. Make 1.5% agarose gel (agarose dissolved in 1× TAE) with ethidium bromide added after heating solution to dissolve agarose.

6. Load samples (~20 μl/samples) into separate wells of the gel. Add a 100 bp DNA ladder to the first well (*see* **Note 4**).

7. Run gel in 1× TAE buffer (voltage will depend on the size of your gel; small gels should be run at a lower voltage than large gels). Running time will depend on the size of the gel and the band separation desired (*see* **Note 5**).

8. Visualize the DNA banding pattern. A gel viewing box with or without a gel camera can be utilized (*see* Fig. 1 and **Note 6**).

4   Notes

1. This assay cannot detect the difference between heterozygote and homozygote PiZ mice.

2. Be sure to fully press down the film as the edges and between the rows of the plate to prevent sample evaporation during the PCR process.

3. Following the PCR step the DNA can sit in the tubes/plate at 4 °C until the electrophoresis gel can be run.

4. If you have a large gel with a large number of samples, you may want to load an additional well with ladder in the middle or last well.

5. If too high a voltage is used, you may see that the center wells run faster than the edge wells, which can make band size

interpretation difficult. When gaining experience running gels, it can be helpful to visualize the progress of the band separation after 30–45 min to see if additional running time is necessary.

6. If the DNA bands are not clear, several steps should be evaluated. If there is little or no band present, additional DNA may need to be loaded into the PCR reaction (and water decreased accordingly) or the DNA may need to be reextracted (this may mean collecting new tissue from the mouse). Poor banding could also be due to primers that have been degraded or contaminated; in that case, a fresh aliquot should be utilized. Additional bands on the gel (small or larger than expected) are not uncommon and if they are common to all samples can be ignored as long as they do not interfere with interpretation of the results. Additional bands can also indicate contamination of the primers, TAE buffer, or mastermix reagent. In that case, try running samples again with fresh reagents.

References

1. Carlson JA, Rogers BB, Sifers RN, Finegold MJ, Clift SM, DeMayo FJ, Bullock DW, Woo SL (1989) Accumulation of PiZ alpha 1-antitrypsin causes liver damage in transgenic mice. J Clin Invest 83(4):1183–1190. doi:10.1172/jci113999

2. Truett GE, Heeger P, Mynatt RL, Truett AA, Walker JA, Warman ML (2000) Preparation of PCR-quality mouse genomic DNA with hot sodium hydroxide and tris (HotSHOT). BioTechniques 29(1):52, 54

Chapter 7

Elastase-Induced Lung Emphysema Models in Mice

Béla Suki, Erzsébet Bartolák-Suki, and Patricia R.M. Rocco

Abstract

Pulmonary emphysema is one of the distinct pathological forms of chronic obstructive pulmonary disease (COPD) that is accompanied by gradual elimination of alveolar tissue, causing reductions in lung recoil and leading to difficulty in breathing. As there is no cure for emphysema, animal models are often used to better understand the pathogenesis and progression of the disease. One widely used animal model of emphysema is the elastase treatment. In this chapter, we describe two methods of elastase-induced emphysema in mice. The first is a single-dose treatment, whereby elastase is introduced oropharengeally into the lung and the structure and/or function of the lungs are studied between 2 days and 4 weeks following the treatment. The second method consists of exposing mice repeatedly (four times) to elastase intratracheally and observing the effects of the treatment 1–4 weeks following the last administration of the enzyme. Both protocols are described in detail, and examples of lung structure and function of the emphysematous mouse lung are provided.

Key words Porcine pancreatic elastase, Anesthesia, Airspace enlargement, Compliance

1 Introduction

Chronic obstructive pulmonary disease (COPD) is a progressive disorder of the respiratory system characterized by chronic bronchitis, involving mostly the small airways, and emphysema, affecting mostly the lung tissue. Emphysema is thought to result mainly from tobacco smoke-induced inflammation, which leads to a loss of elastic tissue, low recoil pressure, flow limitation, and difficulty in breathing. The primary cause of COPD in industrialized countries is tobacco smoking, whereas environmental pollution also contributes to the disease in the developing world [1]. The total number of smoke-related deaths has been estimated to rise from 5.4 million in 2005 to 6.4 million in 2015 and 8.3 million in 2030, and in 2015, COPD is projected to kill 50% more people than HIV/AIDS, accounting for 10% of all deaths worldwide [2]. In addition to smoking, alpha-1 antitrypsin deficiency also leads to emphysema and it has been estimated that there are over 100 million carriers and over a million patients with severe alpha-1 antitrypsin

Florie Borel and Christian Mueller (eds.), *Alpha-1 Antitrypsin Deficiency: Methods and Protocols*, Methods in Molecular Biology, vol. 1639, DOI 10.1007/978-1-4939-7163-3_7, © Springer Science+Business Media LLC 2017

deficiency worldwide making this disease one of the most common hereditary disorders [3].

While the etiology of COPD is known, the corresponding pathogenic mechanisms are not fully understood; hence, no cure is currently available. Within this context, animal models of emphysema have been extensively used to investigate the mechanisms of disease pathogenesis and progression. Specifically, many different mechanisms and signaling pathways involved in emphysema have been identified with the help of various mouse models [4–13]. Furthermore, mouse models of emphysema have also been utilized to test the efficacy of stem cell treatment [14, 15].

As tobacco is the primary cause of emphysema, exposure of mice to cigarette smoke is often used as a model of the human disease [16–19]. In this model, the progression of emphysema invariably ceases once smoke exposure is eliminated, unlike in humans. The pallid mouse has also been used as a model of alpha-1 antitrypsin deficiency [20]. Since in both the cigarette smoke-induced and alpha-1 antitrypsin deficiency-related emphysema, elastolytic activity is maintained over extended periods, induction of emphysema by direct administration of elastase is feasible. Indeed, a known alternative to cigarette smoke exposure is the direct elastolytic injury, the elastase-induced emphysema model, in which the effects of the enzyme on the lung often produces long-term progressive tissue destruction in mice as well as in other species [9, 10, 13, 21, 22]. While the elastase model has limitations, such as differences in apoptosis [23] and remodeling [24] from smoke related emphysema, the elastase model remains useful to investigate the time course of functional and structural changes in lung tissue, as well as the efficacy of novel treatment strategies. The main advantages of the elastase-induced model are that the development of emphysema is both rapid onset and robust and that it remains progressive for many weeks after single- or repeated-dose treatments.

The purpose of this chapter is to describe two models of elastase-induced emphysema in mice: a well-established single-dose treatment and a more recently developed multiple-dose treatment. Following a description of the approaches, we briefly demonstrate the effects of the two treatment strategies.

2 Materials

2.1 Single-Dose Elastase Treatment

1. Plexiglas box for sedation.

2. Isoflurane (2–3%) in oxygen.

3. Rubber band (*see* **Note 1**).

4. 60° inclined platform (*see* **Note 2**).

5. Porcine pancreatic elastase (PPE) (EC134; Elastin Products Company, Inc., Owensville, Missouri, USA) at a concentration of 6 IU PPE/animal (*see* **Note 3**), dissolved in 100 μL phosphate-buffered saline (PBS).

6. Mice ($n = 8$–10); the most commonly used strains are C57BL/6 or BALB/c, which can be purchased, for example, from the Jackson Laboratory (*see* **Note 4**).

2.2 Multiple-Dose Elastase Treatment

1. Plexiglas box for sedation.

2. Sevoflurane (3%) (Abbott, Wiesbaden, Germany).

3. Controlled warming pad.

4. PPE (Sigma Chemical Co., St. Louis, MO, USA) at a concentration of 0.1–0.2 IU PPE/animal (depending on the desired degree of severity), dissolved in 50 μL PBS.

5. 27-gauge tuberculin needle.

6. Lidocaine (2%).

7. 5-0 silk suture.

8. Mice ($n = 8$–10); the most commonly used strains are C57BL/6 or BALB/c.

3 Methods

3.1 Single-Dose Elastase Treatment

1. Once the mice arrive at the animal facility, they should be allowed 3 days to accommodate before treatment. Animals should be fed with laboratory rodent chow, provided water ad libitum, while maintained on a 12-h light–dark cycle. Two to three mice should be housed in a cage, depending on the size of the cage.

2. Following accommodation, prepare the mouse for light sedation by placing the animal in the plexiglas box (*see* **Note 5**) and adding a flow of isoflurane at a rate of between 1.5 and 2 L/min. It takes 1–2 min to sedate the mouse.

3. Once the mouse is sedated, take it out of the box and place it on the inclined platform in the supine position. Carefully open its mouth and attach the rubber band to the upper incisors. Next, hang the rubber band from a nail on the platform (Fig. 1).

4. Using a small pair of forceps (*see* **Note 6**), gently grab the mouse's tongue and pull it out of the mouth to about 0.5–1 cm.

5. Drop the elastase solution onto the distal oropharynx, i.e., the back of the tongue.

6. While holding the tongue with one hand, use the other hand to block both nostrils. Hold the animal in this position for 5–6 s.

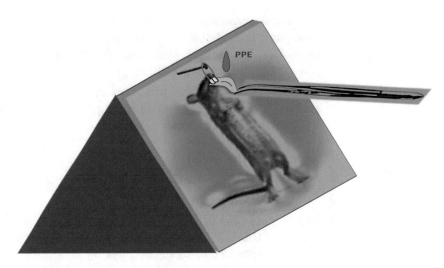

Fig. 1 Schematic of the oropharyngeal PPE treatment of mice. The mouse is lying on an inclined plane and hung from its upper incisors by a red rubber band. The tongue is pulled out by a pair of forceps and PPE (*green*) is dropped to the back of the tongue

With the tongue extended and nose blocked, the animal will be unable to swallow, and the liquid will be aspirated into the lower respiratory tract.

7. Release the mouse's tongue and gently massage the chest for the elastase to flow down and distribute throughout the airways.

8. Remove the rubber band and place the mouse back into the cage.

9. The entire process takes about 2 min and no harm or pain is caused to the animal.

10. The animal should recover in 2–5 min.

11. Watch the animal for an additional 15 min. If the mouse behaves as it did before the procedure, the treatment has been successful.

12. Check on the animal in 4 h, then once a day thereafter for the desired period.

13. Strong, transient inflammation and remodeling occurs by 2 days after the treatment, with both collagen and elastin increasing, which is accompanied by a small but statistically significant airspace enlargement [9] that can be detected by advanced image processing and histological analysis [25]. A stronger response, both in structure and function, can be detected by 3 weeks after treatment. At 4 weeks after treatment, robust emphysema develops.

14. Figure 2 shows representative images of the alveolar structure stained for collagen at 2 days and 3 weeks following saline or

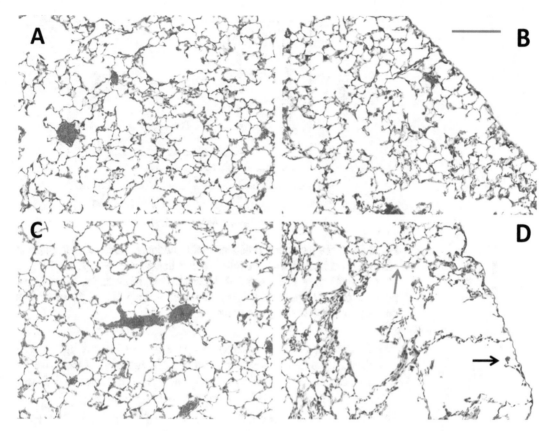

Fig. 2 Examples of collagen-stained (*blue*, Masson's trichrome) images of the lung parenchyma in saline control lungs at Day 2 (**a**) and Day 21 (**c**) and in PPE-treated lungs at Day 2 (**b**) and Day 21 (**d**). The *black arrow* locates a site of alveolar wall rupture with retracted septal wall tissue. The *blue arrow* shows a septal wall under tension containing only collagen. The *red scale bar* in (**b**) represents 100 μm

PPE treatment of mice. Notice the tension in collagen and rupture of alveolar walls which are responsible for airspace enlargement in emphysema [26]. The functional consequences of these structural changes are demonstrated in Fig. 3, showing how dynamic respiratory elastance, the inverse of compliance, measured using the forced oscillation technique at a positive end-expiratory pressure of 3 cmH_2O, gradually decreases with time after PPE treatment.

3.2 Multiple-Dose Treatment

1. Once the mice arrive at the animal facility, they should be allowed 3 days to accommodate before treatment. Animals should be kept under specific pathogen-free conditions.

2. Animals should be fed with laboratory rodent chow, provided water ad libitum, and maintained on a 12-h light–dark cycle. Two or three mice should be housed in a cage, depending on the size of the cage.

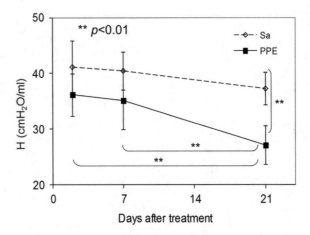

Fig. 3 Elastance parameter H obtained from the forced oscillation technique in saline (Sa: *open circle*) and PPL (*filled square*) treated mice at 2, 7, and 21 days after a single dose PPE treatment. Notice that H, representing the dynamic modulus of the respiratory system, decreases in time and at 21 days, it is different from that in the saline treated mice

3. Following accommodation, prepare the mouse for light anesthesia by placing the animal in the plexiglas box (*see* **Note 5**) and adding a flow of sevoflurane at a rate of between 1.5 and 2 L/min. It takes 1–2 min to sedate the mouse.

4. Once the mouse is sedated, take it out of the box and place it on a controlled heating pad to maintain body temperature constant at 37 °C.

5. Maintain anesthesia with sevoflurane and oxygen. Once the animal is anesthetized, make a 1-cm midline cervical incision to expose the trachea.

6. Instill PPE directly into the trachea with a bent 27-gauge tuberculin needle.

7. Close the cervical incision with 5-0 silk suture, infiltrate lidocaine around the incision to help reduce pain or discomfort caused by surgery and needle puncture, and keep the mouse on oxygen on the controlled heating pad before returning it to the cages (one animal per cage).

8. The entire process takes about 2 min.

9. The animal should recover in 2–3 min.

10. Watch the animal for an additional 30 min. If the mouse behaves as it did before the procedure, the treatment has been successful.

11. Check on the animal in 2 h, then once a day thereafter for the desired period.

12. Repeat **steps 1–11** once a week for 4 weeks.

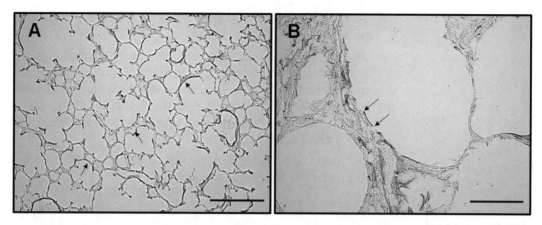

Fig. 4 Elastic fibers in alveolar septa (*black*, Weighert's resorcin fuchsin method with oxidation). (**a**) Control: mice treated with saline, (**b**) Emphysema: animals treated with multiple doses of PPE. Magnification: ×400, scale bar: 50 μm. Note the rupture of elastic fibers in the emphysema group (*arrows*)

13. One week after the last intratracheal instillation, both static and dynamic lung elastance should be reduced, associated with elastic fiber disruption in alveolar septa (Fig. 4), resulting in airspace enlargement, increased neutrophil infiltration, and collagen fiber content (Fig. 4) [15]. Additionally, right ventricle area should be increased and pulmonary artery acceleration time/pulmonary artery ejection time (PAT/PET), an indirect index of pulmonary arterial hypertension, reduced [14].

14. Figure 4 shows representative images of the alveolar structure at 1 week following the last intratracheal instillation of elastase. Note the rupture of elastic fibers (black) inside septal walls.

4 Notes

1. A commercially available color rubber band works well.

2. The inclined platform can be a plexiglas system or it can be built simply from a paper box. A convenient size for the top surface is 5 cm × 15 cm. A nail should be pushed into the top of the incline so that the rubber band can be attached to it. The mouse will be held by the rubber band between the upper incisors and the nail.

3. The amount of PPE in the single-dose method to be delivered to the mouse varies among studies. PPE can also be bought from other vendors and the dose may be slightly different. For example, a dose of 6.75 units of PPE from Sigma-Aldrich (St. Louis, MO) produces similar results to those using PPE from Elastin Products Company, Inc. (Owensville, Missouri, USA).

4. The C57BL/6 mouse is generally more resistant to this treatment than the BALB/c mouse.

5. To place the mouse into the chamber, grasp its tail and pull it, but do not lift the mouse. The mouse will want to escape and move forward. Grab its skin above the neck with your other hand. If you grab the skin deeply, the mouse will be unable to turn its head and bite. This will allow you to hold the mouse firmly and place it in the box.

6. It is useful to pull a small plastic tube over the tip of the forceps to protect the tongue of the mouse.

Acknowledgment

B. Suki and E. Bartolák-Suki were supported by the National Institutes of Health Grant HL-111745. P. Rocco was supported by the Brazilian Council for Scientific and Technological Development (CNPq), the Carlos Chagas Filho Rio de Janeiro State Research Foundation (FAPERJ), and the Department of Science and Technology, Brazilian Ministry of Health (DECIT/MS). P. Rocco would like to express her gratitude to Dr. Milena Vasconcellos for help with microscopy.

References

1. Global strategy for the diagnosis, management and prevention of chronic obstructive pulmonary disease (2013). Accessed Revised 2013

2. Mathers CD, Loncar D (2006) Projections of global mortality and burden of disease from 2002 to 2030. PLoS Med 3 (11):e442. doi:10.1371/journal.pmed.0030442

3. Luisetti M, Seersholm N (2004) Alpha1-antitrypsin deficiency. 1: epidemiology of alpha1-antitrypsin deficiency. Thorax 59 (2):164–169

4. Cantor JO, Cerreta JM, Ochoa M, Ma S, Chow T, Grunig G, Turino GM (2005) Aerosolized hyaluronan limits airspace enlargement in a mouse model of cigarette smoke-induced pulmonary emphysema. Exp Lung Res 31 (4):417–430

5. de Santi MM, Martorana PA, Cavarra E, Lungarella G (1995) Pallid mice with genetic emphysema. Neutrophil elastase burden and elastin loss occur without alteration in the bronchoalveolar lavage cell population. Lab Investig 73(1):40–47

6. Fisk DE, Kuhn C (1976) Emphysema-like changes in the lungs of the blotchy mouse. Am Rev Respir Dis 113(6):787–797

7. Foronjy RF, Mercer BA, Maxfield MW, Powell CA, D'Armiento J, Okada Y (2005) Structural emphysema does not correlate with lung compliance: lessons from the mouse smoking model. Exp Lung Res 31(6):547–562

8. Gardi C, Martorana PA, Calzoni P, van Even P, de Santi MM, Cavarra E, Lungarella G (1992) Lung collagen synthesis and deposition in tight-skin mice with genetic emphysema. Exp Mol Pathol 56(2):163–172

9. Hamakawa H, Bartolak-Suki E, Parameswaran H, Majumdar A, Lutchen KR, Suki B (2011) Structure-function relations in an elastase-induced mouse model of emphysema. Am J Respir Cell Mol Biol. doi:10.1165/rcmb. 2010-0473OC. 2010-0473OC [pii]

10. Lucattelli M, Cavarra E, de Santi MM, Tetley TD, Martorana PA, Lungarella G (2003) Collagen phagocytosis by lung alveolar macrophages in animal models of emphysema. Eur Respir J 22(5):728–734

11. O'Donnell MD, O'Connor CM, FitzGerald MX, Lungarella G, Cavarra E, Martorana PA (1999) Ultrastructure of lung elastin and collagen in mouse models of spontaneous emphysema. Matrix Biol 18(4):357–360

12. Shiomi T, Okada Y, Foronjy R, Schiltz J, Jaenish R, Krane S, D'Armiento J (2003) Emphysematous changes are caused by degradation of type III collagen in transgenic mice expressing MMP-1. Exp Lung Res 29(1):1–15

13. Yao H, Arunachalam G, Hwang JW, Chung S, Sundar IK, Kinnula VL, Crapo JD, Rahman I (2010) Extracellular superoxide dismutase protects against pulmonary emphysema by attenuating oxidative fragmentation of ECM. Proc Natl Acad Sci U S A 107(35):15571–15576. doi:10.1073/pnas.1007625107

14. Antunes MA, Abreu SC, Cruz FF, Teixeira AC, Lopes-Pacheco M, Bandeira E, Olsen PC, Diaz BL, Takyia CM, Freitas IP, Rocha NN, Capelozzi VL, Xisto DG, Weiss DJ, Morales MM, Rocco PR (2014) Effects of different mesenchymal stromal cell sources and delivery routes in experimental emphysema. Respir Res 15:118. doi:10.1186/s12931-014-0118-x

15. Cruz FF, Antunes MA, Abreu SC, Fujisaki LC, Silva JD, Xisto DG, Maron-Gutierrez T, Ornellas DS, Sa VK, Rocha NN, Capelozzi VL, Morales MM, Rocco PR (2012) Protective effects of bone marrow mononuclear cell therapy on lung and heart in an elastase-induced emphysema model. Respir Physiol Neurobiol 182 (1):26–36. doi:10.1016/j.resp.2012.01.002

16. Bracke KR, Dentener MA, Papakonstantinou E, Vernooy JH, Demoor T, Pauwels NS, Cleutjens J, van Suylen RJ, Joos GF, Brusselle GG, Wouters EF (2010) Enhanced deposition of low-molecular-weight hyaluronan in lungs of cigarette smoke-exposed mice. Am J Respir Cell Mol Biol 42(6):753–761. doi:10.1165/rcmb.2008-0424OC

17. Cavarra E, Bartalesi B, Lucattelli M, Fineschi S, Lunghi B, Gambelli F, Ortiz LA, Martorana PA, Lungarella G (2001) Effects of cigarette smoke in mice with different levels of alpha (1)-proteinase inhibitor and sensitivity to oxidants. Am J Respir Crit Care Med 164 (5):886–890. doi:10.1164/ajrccm.164.5.2010032

18. Hautamaki RD, Kobayashi DK, Senior RM, Shapiro SD (1997) Requirement for macrophage elastase for cigarette smoke-induced emphysema in mice. Science 277 (5334):2002–2004

19. Ofulue AF, Ko M, Abboud RT (1998) Time course of neutrophil and macrophage elastinolytic activities in cigarette smoke-induced emphysema. Am J Phys 275(6 Pt 1): L1134–L1144

20. Martorana PA, Brand T, Gardi C, van Even P, de Santi MM, Calzoni P, Marcolongo P, Lungarella G (1993) The pallid mouse. A model of genetic alpha 1-antitrypsin deficiency. Lab Investig 68(2):233–241

21. Hantos Z, Adamicza A, Janosi TZ, Szabari MV, Tolnai J, Suki B (2008) Lung volumes and respiratory mechanics in elastase-induced emphysema in mice. J Appl Physiol 105 (6):1864–1872. doi:10.1152/japplphysiol. 90924.2008. 90924.2008 [pii]

22. Ito S, Ingenito EP, Brewer KK, Black LD, Parameswaran H, Lutchen KR, Suki B (2005) Mechanics, nonlinearity, and failure strength of lung tissue in a mouse model of emphysema: possible role of collagen remodeling. J Appl Physiol 98:503–511

23. Sawada M, Ohno Y, La BL, Funaguchi N, Asai T, Yuhgetsu H, Takemura G, Minatoguchi S, Fujiwara H, Fujiwara T (2007) The Fas/Fas-ligand pathway does not mediate the apoptosis in elastase-induced emphysema in mice. Exp Lung Res 33(6):277–288. doi:10.1080/01902140701509458

24. Takahashi A, Majumdar A, Parameswaran H, Bartolak-Suki E, Suki B (2014) Proteoglycans maintain lung stability in an elastase-treated mouse model of emphysema. Am J Respir Cell Mol Biol 51(1):26–33. doi:10.1165/rcmb.2013-0179OC

25. Parameswaran H, Majumdar A, Ito S, Alencar AM, Suki B (2006) Quantitative characterization of airspace enlargement in emphysema. J Appl Physiol 100(1):186–193

26. Kononov S, Brewer K, Sakai H, Cavalcante FS, Sabayanagam CR, Ingenito EP, Suki B (2001) Roles of mechanical forces and collagen failure in the development of elastase-induced emphysema. Am J Respir Crit Care Med 164(10 Pt 1):1920–1926

Chapter 8

Assessing Structure–Function Relations in Mice Using the Forced Oscillation Technique and Quantitative Histology

Harikrishnan Parameswaran and Béla Suki

Abstract

The structure and function of the lung gradually becomes compromised during the progression of emphysema. In this chapter, we first describe how to assess and evaluate lung function using the forced oscillation technique. Next, we provide details on how to use the Flexivent system to measure respiratory mechanical parameters in mice. We also describe the outlines of how to set up a homemade forced oscillatory system and use it to measure respiratory mechanics. To characterize the structure from standard histological images, we describe a method that is highly sensitive to early emphysema. Correlating structural information such as equivalent alveolar diameter and its variance with respiratory elastance or compliance, provides structure–function relationships that can subsequently reveal novel mechanisms of emphysema progression or be used to track the effectiveness of treatment.

Key words Emphysema, Lung elastance, Compliance, Airspace enlargement, Alveolar diameter

1 Introduction

Emphysema is a subtype of chronic obstructive pulmonary disease (COPD) that is also described as a progressive disorder of the lung characterized by the gradual and irreversible destruction of alveolar walls, which leads to abnormally enlarged airspaces and reduced surface area for gas exchange [1–3]. While tobacco smoke is believed to be the primary risk factor for emphysema [4], other factors such as environmental pollutants [5], senescence [6, 7], nutrition [8, 9], and genetic predispositions [10] may also cause emphysema. Each of these risk factors triggers a series of interconnected biochemical processes that lead to cell death and the degradation of protein fibers that make up the alveolar walls. Consequently, the alveolar walls rupture and abnormally enlarged airspaces appear [11, 12]. Over time, the destruction of the alveolar walls become progressive and functional consequences follow [13].

Florie Borel and Christian Mueller (eds.), *Alpha-1 Antitrypsin Deficiency: Methods and Protocols*, Methods in Molecular Biology, vol. 1639, DOI 10.1007/978-1-4939-7163-3_8, © Springer Science+Business Media LLC 2017

Emphysema is defined in terms of the destruction of alveolar septa and the appearance of abnormally enlarged airspaces. Quantifying microscopic structural changes in lung tissue may be the best way to characterize early changes following the onset of the disease. It is therefore, important to be able to link our understanding of microscopic structural changes to macroscopic changes in lung function.

The forced oscillation technique (FOT) is the best method to evaluate the functional properties of the lung and the respiratory system [14]. It is based on applying prescribed pressure variations in a desired frequency range at the airway opening and then measuring the response of the lung or respiratory system, which reflects the flow into the lung. The complex ratio of the pressure and the flow in the frequency domain is the impedance that can be evaluated in relation to the frequencies that were used in the input signals. For efficient measurements, the input is often designed to contain multiple frequencies. That is, it is a sum of sine waves with specific magnitudes, frequencies, and phases. The real part of the impedance, the component of the normalized pressure in phase with the flow, is called the resistance. Accordingly, the imaginary part, which is the out of phase component, contains information on tissue stiffness and airway inertance. When the impedance is obtained over a wide enough frequency range that encompasses the spontaneous breathing frequency, features of the impedance can be fitted to lumped parameter models. The most widely used model is the constant phase model of tissue impedance that is complemented with an airway compartment [15]. The model parameters such as airway resistance as well as tissue elastance or compliance can be plotted against the positive end-expiratory pressure (PEEP) [16] or some structural parameters such as alveolar diameter or its heterogeneity [17]. When regional tissue structure is separately assessed, additional interesting structure–function relations can also be formulated which may provide insight into the spatial progression of the disease [18].

The morphometric index that is most commonly used for quantifying emphysema at microscopic scales is the mean linear intercept (L_m) [19, 20]. The L_m has its theoretical basis in the Cauchy–Crofton formula [21, 22], which relates the mean chord length of a three dimensional object to its volume (V) and surface area (S) as: $L_m = 4\,V/S$. Airspace chord lengths are measured from two dimensional histological sections of the lung and assuming that these measurements represent the true three dimensional chord lengths, the surface area to volume ratio can be estimated based on a modified version of the Cauchy–Crofton equation [19]. However, in diseases with spatially heterogeneous structure such as emphysema, it is unclear if the above assumption is valid. More importantly, several studies have questioned the sensitivity of this index to detect early emphysema [23–25]. For example, it was found that the L_m is normal in 32% of emphysematous patients

[25]. In an elastase-induced model of emphysema in rats, it was reported that the L_m was not very sensitive to the structural changes even after 9 weeks following treatment [26]. Additionally, it is worth noting that the airspace wall per unit volume $(2/L_m,)$, which is an index derived from the L_m, is only abnormal in 26% of surgically resected patients with severe emphysema [27].

In order to link the structural and functional changes in emphysema, we need to look beyond the mean linear intercept technique. In 1994, Saetta et al. [28] found that when human emphysema patients were classified into four categories based on patterns observed in two-dimensional (2D) histological images ranging from a very homogeneous destruction pattern to a highly heterogeneous pattern, the subjects showed a significant difference in static compliance. However, the total destruction quantified by the L_m was similar among the groups, suggesting that apart from the amount of tissue lost, patterns observed at the microscale should have a strong influence on macroscale function. Indeed, this has been validated in a three-dimensional (3D) computational model of emphysema [13].

In this chapter, we first introduce two methods of FOT to characterize lung and respiratory function in mice. The easier method is an FOT built into a commercially available rodent mechanical ventilator system, whereas the second method provides details on how to build an FOT system in the laboratory. Next, we describe an automated method for quantifying parenchymal destruction that accounts for the heterogeneous patterns of tissue loss in emphysema [23]. The method produces an area weighted mean diameter index (D_2) that has been shown to be more sensitive for detecting subtle airspace enlargements than traditional techniques such as the L_m. Finally, we demonstrate that by using the FOT and quantitative histology, we can link microscopic patterns of tissue destruction to changes in lung function characterized by respiratory system compliance.

2 Materials

2.1 FOT Measurements Using the FlexiVent System

1. A fully functional FlexiVent system (Scireq, Montreal, Canada) to ventilate the mice.
2. The FlexiVent system includes a programmable ventilator that can generate pressure waveforms required for the FOT measurements.

2.2 Homemade FOT System

1. A rodent mechanical ventilator such as the Harvard ventilator.
2. An airtight loudspeaker-in-box system which is capable of producing pressure oscillations at 2 Hz.

3. Various tubes including a T-tube and a three-way stopcock that can be attached to the outlet of the mechanical ventilator as well as each other.

4. A power amplifier for the loudspeaker.

5. A pressure transducer with a range of at least ± 30 cm H_2O, but not more than ± 100 cm H_2O (e.g., Omega, mini solid state pressure sensor PXSCX-001DV) and a flowmeter that can measure small flows (e.g., Fleisch 00000 pneumotachograph).

6. An analogue low-pass filter with at least two channels (at least a fourth order with a cutoff around 30 Hz).

7. An additional low-pass filter for the output signal (this may be part of a three-channel low pass filer in #6 and at least a fourth order with a cutoff around 30 Hz).

8. A two-channel amplifier.

9. An AD/DA converter.

10. A laptop or desktop computer.

11. Software for the AD/DA converter (e.g., LabView).

12. Software such as Matlab to create input signals and compute the power spectra.

2.3 Image Processing

1. Digital images of standard histological sections of lungs fixed at Total Lung Capacity (TLC). Typically, hematoxylin and eosin-stained sections are used for automated processing. However, all that is required are images where the tissue and airspaces can be clearly differentiated in grayscale.

2. A digital image processing software platform. Any software that is capable of reading and analyzing modern imaging formats can be used. However, platforms such as Matlab (Mathworks, Natick, MA), NIH ImageJ, or Python are recommended because of their built-in image processing libraries.

3. A digital computer capable of running any of the abovementioned image processing software.

3 Methods

3.1 FOT Measurements Using the Flexivent System

1. Start the Flexivent software and carry out the static calibrations followed by the dynamic calibrations according to the manufacturer's instructions using the noninteger multiple perturbation signal and the pressure–volume signal. These signals need to be loaded into the Flexivent software.

2. Prepare the mouse by properly anesthetizing it and then carefully tracheotomize the animal.

3. Set the default ventilation setting between 180 and 240 breaths per minute and 8 ml/kg tidal volume, and then immediately start the ventilation.

4. Connect the tracheostomy tube to the Flexivent piston, which also ventilates the animal.

5. Placed the animal on a low PEEP of 2 or 3 cm H_2O (*see* **Note 1**).

6. In order to standardize the volume history before the FOT measurement, initiate a pressure–volume curve measurement and ventilate the mouse for 1 min using the default ventilation mode.

7. Initiate the FOT measurement by clicking on the perturbation signal in the memory setting.

8. Ventilate the animal for about 20 s, deliver the pressure–volume curve, ventilate for one additional minute, and deliver the FOT. Repeat these steps at least three more times and this completes the FOT data acquisition (*see* **Note 2**).

9. The software automatically fits the data with a model called the constant-phase model of the respiratory system. This model includes a constant resistance (R), an inertance (I), a tissue damping parameter (G), and a tissue elastance parameter (H). The fitting of the model to the impedance data effectively partitions the respiratory system impedance to the airway and tissue compartments. In emphysema, it is the tissue compartment that is of primary interest. Download these parameters from the Flexivent software, compute the compliance as $C = 1/H$, and average their values. If one value is a significant outlier, drop that value (*see* **Note 3**).

10. To make sure the Flexivent measures correctly, physical models of the respiratory system can be used (*see* **Note 4**). The inverse of the compliance of a plastic chamber should match the H value that is obtained from Flexivent within 5%.

11. If needed, change the PEEP and repeat the procedure (*see* **Note 5**).

3.2 Homemade FOT Measurements

1. Set up the FOT system as follows (Fig. 1).

2. Connect the output of the AD/DA converter to the computer and connect the DA port to the input of one of the low-pass filters. Connect the output of this low-pass filter to the power amplifier that is attached to the loudspeaker.

3. Attach the flowmeter to the outlet of the loudspeaker-inbox through a tube and attach a T-tube to the flowmeter. Connect the straight path of the T-tube to the three-way stopcock, which should then be attached to tracheostomy tube. The path from the loudspeaker to the tracheostomy tube should

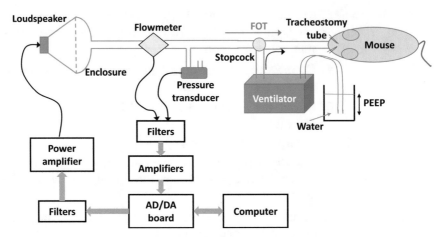

Fig. 1 A setup for the forced oscillation technique (FOT). An anesthetized and tracheotomized mouse is connected to a commercial rodent mechanical ventilator via a stopcock. The other end of the stopcock is attached the FOT setup, which consists of a loudspeaker with an enclosed front that is able to deliver pressure waves through a flowmeter into the mouse. A pressure transducer measures airway opening pressure. The flow and pressure are passed through an analogue filter and amplifier and sampled by an AD/DA board and a computer. A pseudorandom noise signal is sent through the D/A converter, filtered and power amplified and this signal drives the loudspeaker

be straight. The pressure transducer should be attached to the middle port of the T-tube and the middle branch of the stopcock should be attached to the mechanical ventilator.

4. Connect the electrical output of the flowmeter to the input of one of the low-pass filters. Next, connect the output of the low-pass filter to the amplifier and then connect the amplifier to the AD board.

5. Repeat the previous steps with the pressure transducer and this completes the assembly of the setup.

6. Using software such as Matlab, create a pseudorandom noise signal consisting of a sum of 8–10 sine waves with the frequencies covering the range of 1–20 Hz (*see* **Note 6**). The sampling rate should be at least 3–4 times higher than the highest frequency of the sum-of-sines signal.

7. Choose the cutoff frequency of the low-pass filter so that your signals will not have significant energy above the Nyquist frequency (half the sampling rate, *see* **Note 7**).

8. Perform an absolute calibration of the pressure transducer and the flowmeter (*see* **Note 8**).

9. To calibrate the system, use a small glass chamber with a known compliance such as a small rigid chamber (*see* **Note 4**). After making sure that there is no leakage in the system, start simultaneously sending out the pseudorandom noise, sample pressure, and flow. Set the power amplifier so that the peak-to-peak pressure amplitude is around 2–3 cm H_2O.

10. Evaluate the results in the frequency domain by using Fourier analysis (*see* **Note 9**).

11. Once the agreement between the theoretical compliance of the chamber and the measured compliance is satisfactory, the system is ready to measure the mechanical impedance of the mouse.

12. Following the proper anesthesia and tracheostomy procedures, connect the mouse to the mechanical ventilator and ventilate the animal as described in Subheading 3.1, **steps 3–5**.

13. Follow the protocol starting at Subheading 3.1, **step 8** in order to obtain FOT-based respiratory impedance of the mouse. Use the stopcock to switch between mechanical ventilation and FOT measurements. If the ventilator is not equipped with a PEEP setting, then place the outlet of the ventilator into a water trap by lowering the outlet tube 2–3 cm underwater. To achieve similar volume standardizations as the Flexivent, stack three inspirations on each other by blocking the outlet for three breathing cycles.

14. Compute the impedance of the respiratory system (*see* **Note 9**). An estimate of the tissue elastance parameter H is obtained as the inverse of the frequency dependent compliance at 2 Hz (*see* **Note 10**). Compute the average of C or H and plot it against PEEP or an index of the alveolar structure as described below in Subheading 3.3.

3.3 Evaluation of Area Weighted Mean Diameter

1. The general steps are outlined in Fig. 2. The image is first converted from grayscale to a binary image where a value of 1 represents airspaces and a value of 0 represents tissue (*see* **Notes 11** and **12**).

2. The binary image is used to calculate a distance map where the value of each pixel is replaced by its distance to the nearest tissue pixel multiplied by -1. This step produces an image with multiple local minima, where each minima is created by the curvature of the surrounding alveolar septa.

3. Minima corresponding to the minor irregularities lead to very shallow local minima, which can be removed by using the H-minima transformation. This suppresses all minima in the image with values less than the specified threshold.

4. The deeper minima in the distance map are retained, and are then used as markers for the watershed transformation. Intuitively, the watershed transformation can be described as follows: if we picture the distance map as a 3D surface, the watershed transform is analogous to flooding the 3D surface by placing a water source in each of the identified minima, in order to flood the entire 3D surface of the distance map from these sources. Where water from two different sources meet, lines are

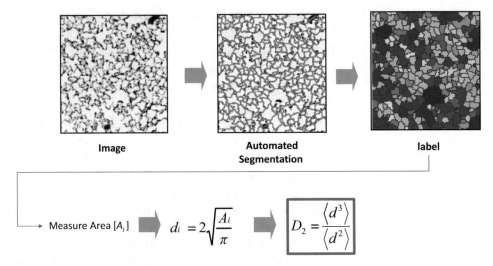

Fig. 2 Steps of image processing. A grayscale image of the lung tissue is thresholded and converted into a binary image on which a distance map is created. The watershed transform is then used to segment the image into separate airspaces. From the area of each airspace (A_i), the equivalent diameter d_i is computed. Finally, from the set of d_i values, the area weighted mean diameter D_2 is obtained. Reprinted with permission of the American Thoracic Society. Copyright © 2016 American Thoracic Society. Hamakawa et al. (2011). The American Journal of Respiratory Cell and Molecular Biology is an official journal of the American Thoracic Society

drawn which separate the corresponding image into distinctly labeled regions, each of which corresponds to a single alveolar airspace. An example of original and processed images is shown in Fig. 2.

5. After defining the airspaces, vessels and airways are excluded manually from the analysis (*see* **Note 13**) and the area of each airspace (A) can be measured by counting the number of pixels enclosed by each airspace (*see* **Note 14**).

6. An equivalent diameter (*d*) of an airspace with area A is then calculated as:

$$d = 2\sqrt{\frac{A}{\pi}}.$$

7. From all of the airspaces measured for one lung, the area weighted mean diameter D_2 can be calculated as: $D_2 = \frac{\langle d^3 \rangle}{\langle d^2 \rangle}$ where $< \ldots >$ denotes taking the average over all values.

3.4 Structure–Function Relations

1. Once FOT parameters such as respiratory system compliance and resistance are obtained in several groups of mice that had received some form of treatment that causes emphysema, these functional parameters should be correlated with structural

parameters such as mean airspace diameter, variance of airspace diameter, and D_2.

2. If there is sufficient data for both the function and structure from the same mice, regression analysis can be performed between these variables. Figure 3 demonstrates that respiratory system compliance (*see* **Note 15**) as well as its standard

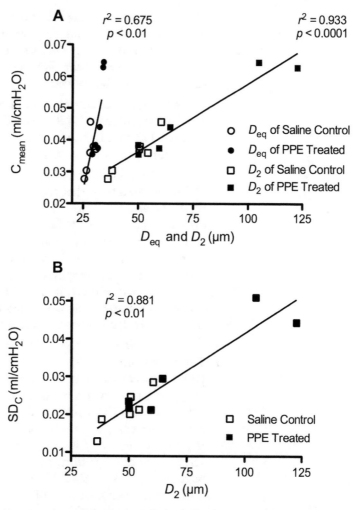

Fig. 3 Structure–function relations in the porcine pancreatic elastase treated mouse model of pulmonary emphysema. Impedance data were fitted by the constant phase model complemented with a distribution of compliances. (**a**) The mean compliance C_{mean} correlates well with the mean equivalent diameter (D_{eq}), as well as the mean area weighted diameter D_2. Notice, however, that function correlates much better with D_2 which incorporates the effects of both general airspace enlargement and the heterogeneity of it. (**b**) The standard deviation of compliance (SD_c) also correlates with D_2. Reprinted with permission of the American Thoracic Society. Copyright © 2016 American Thoracic Society. Hamakawa et al. (2011). The American Journal of Respiratory Cell and Molecular Biology is an official journal of the American Thoracic Society

deviation (*see* **Note 16**) correlates well with the structural heterogeneity as characterized by D_2. These types of results provide insight into how changes in structure alter function in emphysema.

4   Notes

1. On older versions of the FlexiVent, this can be done by attaching a tube to the expiratory port of the FlexiVent and placing the other end of the tube in water to a depth of 2 or 3 cm. For the newer versions of the FlexiVent (version 7 and above), this can be set from the software.

2. Oftentimes, more information can be obtained if the FOT data acquisition is repeated at 3–4 PEEP levels such as 0, 3, 6, and 9 (or even 12) cm H_2O. In this case, a graph of H or compliance C as a function of PEEP is useful [16].

3. The normal value of the H in a 12-week mouse is about 30 cm H_2O/ml. Assuming that the average is near 30 cm H_2O, if a single H value is below 20 or above 40, that value can be dropped because the coefficient of variation of H is not larger than 25%. In an emphysematous mouse, the value of H can be anywhere between 5 and 25 cm H_2O depending on the type and severity of the emphysema model [17, 18]. If there are more than two outliers, it is best to reexamine the anesthesia because the animal may have been breathing during the measurement. If that is the case, add a supplemental dose of anesthetics and repeat the entire FOT protocol.

4. The compliance of a rigid chamber can be estimated as follows: If the heat conductivity of the chamber wall is not very high (e.g., a plastic or wood container), the changes in pressure and volume due to the forced oscillations can be taken as isothermal. In this case, Boyle's law says that the product of pressure (P) and volume (V) is constant: $P \times V = \text{const.}$ Take the variation of the equation, $\Delta P \times V + P \times \Delta V = 0$, which can be rearranged as $-\Delta V / \Delta P = V/P$. Note that the left hand side is exactly the compliance C of the chamber. Thus, C is just the volume divided by ambient temperature. For the measurement purposes, we can use the normal mouse lung with a $C = 0.03$ ml/cm H_2O, which gives a volume of 31.2 ml for the chamber, assuming atmospheric pressure is 1040 cm H_2O. Therefore, a 31 ml plastic chamber should be a good physical model of the mouse lung or respiratory system. If the heat loss is significant, a polytrophic (transition between isothermal and adiabatic) process occurs in the chamber and this provides a slightly lower compliance that decreases with frequency and a real part that decreases nearly hyperbolically with the

frequency toward zero and with a magnitude that is about 5–7% of the imaginary part of the impedance.

5. Usually, between 0 and 3 and maybe even 6 cm H_2O of PEEP, H slightly decreases in the normal animal and then plateaus or even starts increasing at 9 and 12 cm H_2O of PEEP. In the case of an emphysematous mouse, the level of H is lower at PEEPs of 0, 3, and 6 cm H_2O. Again, depending on the type and severity of emphysema, H may start to increase at PEEPs of 9 and 12 cm H_2O with values that may be larger than those in the normal mouse [29].

6. The best combinations of frequencies are those which minimize the nonlinear interactions as this provides smooth estimates of the impedance. A simple choice is a series of prime numbers excluding 1 such as 2, 3, 5, 7, 11, 13, 17, and 19. These can be also the exact frequencies (*see* **Note 7**). The magnitudes can decrease hyperbolically from 1 to 5–6 Hz and then be constant, whereas the phase angles can be randomly selected. It is also possible to use even more selective waveforms that minimize higher order nonlinearities [30].

7. The following can be used as a proper choice. Set the frequencies exactly as in **Note 5** (e.g., 2, 3, 5, 7, 11, 13, 17, and 19 Hz). Set the length of the fast Fourier transform (FFT) to 128 as well as the sampling and shift rates as 128 Hz. These settings will result in a frequency resolution of 1 Hz and the time period of the signal that is 1 s. Copy the signal three times after the original in order to get a total signal length of 512 points, which corresponds to a 4 s long perturbation signal. Use low-pass filtering with a cutoff frequency between 25 and 40 Hz with an 8th-order filter to make sure no energy is in the input and output signals above 64 Hz.

8. Make sure that the electrical output of the pressure transducer shows zero volts when both inlets are open to the atmosphere. To calibrate the pressure transducer, use a water manometer. Connect one inlet of the transducer to the manometer and leave the other open to the atmosphere. Create a 10 cm height difference between the water levels and record the electrical output, increase the height difference to 20 cm then to 30 and 40 cm and record the outputs. A linear regression between electrical output and height difference will give you the proper calibration constant. To calibrate the flowmeter, connect a 10 ml syringe to the flowmeter. Start recording the output of the flowmeter and then slowly push the 10 ml air through the flowmeter. Keep recording for a few more seconds. Using your software (e.g., Matlab), remove the constant at the beginning (before pushing through the air) from the electrical signal and then integrate the signal. The numerical value is equivalent to 10 ml and this is your calibration coefficient. During

impedance measurements, convert the electrical signals of pressure and flow to the calibrated pressure and flow.

9. To evaluate the impedance, carry out FFTs as follows [14, 15, 31]. Multiply the recorded pressure and flow data with the proper calibration constants. Isolate the first Fourier block of pressure and flow data (128 points if you have followed **Note** 7) and apply an FFT on both signals. Compute the auto- and cross-power spectra of flow and pressure by multiplying, at each frequency, the complex flow spectra with its complex conjugate and the pressure with the complex conjugate of the flow. Move your window by 50% of the FFT length (e.g., 64 points), repeat the calculations and add the auto- and cross-power spectra to those obtained in the previous step. Move your window and repeat the process of adding the spectra until you reach the end of the records as this should give you seven blocks of data. At the end of this procedure, divide the cross-power spectra with the auto-power spectra at the frequencies of interest (e.g., if you have used **Note** 7 setting, then do the calculations at the 3rd, 4th, ..., array elements since the first corresponds to the DC component). More advanced computations can be obtained by increasing the overlap between windows, using a Hamming window and removing the trends from the signal. Finally, as for quality check, one can compute the coherence function at every frequency of interest as defined as the magnitude square of the cross-power spectrum normalized by the product of the auto-power spectra of pressure and flow.

10. The H parameter can be obtained at the lowest frequency of the impedance Z as $H = -\text{Imag}(Z) \times 2 \times \pi \times 2(\text{Hz})$. This is not exact, but may be a very close approximation. If a better estimation is needed, the so-called constant phase tissue impedance in series with an airway compartment must be formally fitted to the measured Z. The model is given by $Z = R + j \times I + (G - j \times H)/\omega^{\alpha}$, where j is the imaginary unit, ω is the circular frequency, and $\alpha = (2/\pi) \times \text{atan}(H/G)$ [15]. If more advanced modeling is needed, one can fit a distributed model to the impedance spectra that estimates the distribution of H [32] which can then be further correlated with structural parameters [18].

11. Thresholding the grayscale image to obtain a binary image sometimes produces large sections of completely black pixels toward the edges of the image. This is caused by the nonuniform illumination of the slide under the light microscope. This can be fixed during imaging by applying a flat field correction in the imaging software. If such an option is not available, then the thresholding can be done by splitting up the image into four nonoverlapping blocks in each dimension and

calculating the threshold automatically using a suitable algorithm such as the Otsu's method of variances.

12. The most sensitive parameter in the segmentation process is the threshold for the H-minima transformation that is used to remove shallow minima (as described in Subheading 3.3). Too high of a threshold will result in under segmentation and too low of a threshold will result in oversegmentation.

13. It is recommended to split the implementation in two. The first step in the workflow is an automated program and it runs overnight processing over a large library of images. This step saves the output of the watershed segmentation program in a digital format. The second step is manual and includes a graphical user interface which overlays the output of the segmentation over the original image. An experienced user should then remove the large airways and blood vessels with a single click on that region.

14. Image size greatly affects the speed of processing. A maximum image size of 1024 × 1024 pixels is recommended.

15. In mice, the chest wall is very soft and measurements of respiratory compliance are a reasonable surrogate of lung compliance [33].

16. Advanced structure–function relations can also be generated as follows: The lung impedance spectra in diseases are invariably influenced by mechanical heterogeneities such as time constant inequalities. Ito et al. 2004 developed a model in which the regional compliances of the lung are distributed over a given range that generates a distribution of times constants [32]. By fitting the analytical form of this model to the measured impedance data, it is possible to estimate both the mean compliance (C_{mean}) and the standard deviation of compliances (SD_C) in the lung. The SD_C can then be plotted against measures of structural heterogeneity as in Fig. 3b.

Acknowledgment

B. Suki was supported by the National Institutes of Health Grants HL-111745 and H. Parameswaran was supported by HL129468.

References

1. Snider G, Kleinerman J, Thurlbeck W, Bengali Z (1985) The definition of emphysema. Report of a national heart, lung, and blood institute, division of lung diseases workshop. Am Rev Respir Dis 132:182–185

2. Stockley RA, Mannino D, Barnes PJ (2009) Burden and pathogenesis of chronic obstructive pulmonary disease. Proc Am Thorac Soc 6 (6):524–526. doi:10.1513/pats.200904-016DS. 6/6/524 [pii]

3. Tuder RM, McGrath S, Neptune E (2003) The pathobiological mechanisms of emphysema models: what do they have in common? Pulm Pharmacol Ther 16(2):67–78

4. Celli BR, MacNee W, Agusti A, Anzueto A, Berg B, Buist AS, Calverley PMA, Chavannes N, Dillard T, Fahy B, Fein A, Heffner J, Lareau S, Meek P, Martinez F, McNicholas W, Muris J, Austegard E, Pauwels R, Rennard S, Rossi A, Siafakas N, Tiep B, Vestbo J, Wouters E, ZuWallack R (2004) Standards for the diagnosis and treatment of patients with COPD: a summary of the ATS/ERS position paper. Eur Respir J 23:932–946. doi:10.1183/09031936.04.00014304

5. Girod CE, King TEJ (2005) COPD: a dust-induced disease? Chest 128:3055–3064. doi:10.1378/chest.128.4.3055

6. Pauwels RA, Rabe KF (2004) Burden and clinical features of chronic obstructive pulmonary disease (COPD). Lancet 364:613–620. doi:10.1016/S0140-6736(04)16855-4

7. Tsuji T, Aoshiba K, Nagai A (2006) Alveolar cell senescence in patients with pulmonary emphysema. Am J Respir Crit Care Med 174:886–893. doi:10.1164/rccm.200509-1374OC

8. Bishai JM, Mitzner W (2008) Effect of severe calorie restriction on the lung in two strains of mice. Am J Physiol Lung Cell Mol Physiol 295:L356–L362. doi:10.1152/ajplung.00514.2007

9. Karlinsky JB, Goldstein RH, Ojserkis B, Snider GL (1986) Lung mechanics and connective tissue levels in starvation-induced emphysema in hamsters. Am J Phys 251:R282–R288

10. Ganrot PO, Laurell CB, Eriksson S (1967) Obstructive lung disease and trypsin inhibitors in alpha-1-antitrypsin deficiency. Scand J Clin Lab Invest 19:205–208

11. Kononov S, Brewer K, Sakai H, Cavalcante FS, Sabayanagam CR, Ingenito EP, Suki B (2001) Roles of mechanical forces and collagen failure in the development of elastase-induced emphysema. Am J Respir Crit Care Med 164(10 Pt 1):1920–1926

12. Suki B, Lutchen KR, Ingenito EP (2003) On the progressive nature of emphysema: roles of proteases, inflammation, and mechanical forces. Am J Respir Crit Care Med 168(5):516–521

13. Parameswaran H, Majumdar A, Suki B (2011) Linking microscopic spatial patterns of tissue destruction in emphysema to macroscopic decline in stiffness using a 3D computational model. PLoS Comput Biol 7(4):e1001125. doi:10.1371/journal.pcbi.1001125. ARTN e1001125

14. Bates JH, Irvin CG, Farre R, Hantos Z (2011) Oscillation mechanics of the respiratory system. Compr Physiol 1(3):1233–1272. doi:10.1002/cphy.c100058

15. Hantos Z, Daroczy B, Suki B, Nagy S, Fredberg JJ (1992) Input impedance and peripheral inhomogeneity of dog lungs. J Appl Physiol 72(1):168–178

16. Brewer KK, Sakai H, Alencar AM, Majumdar A, Arold SP, Lutchen KR, Ingenito EP, Suki B (2003) Lung and alveolar wall elastic and hysteretic behavior in rats: effects of in vivo elastase treatment. J Appl Physiol 95(5):1926–1936

17. Hamakawa H, Bartolak-Suki E, Parameswaran H, Majumdar A, Lutchen KR, Suki B (2011) Structure-function relations in an elastase-induced mouse model of emphysema. Am J Respir Cell Mol Biol. doi:10.1165/rcmb.2010-0473OC. 2010-0473OC [pii]

18. Sato S, Bartolak-Suki E, Parameswaran H, Hamakawa H, Suki B (2015) Scale dependence of structure-function relationship in the emphysematous mouse lung. Front Physiol 6:146. doi:10.3389/fphys.2015.00146

19. Dunnill M (1962) Quantitative methods in the study of pulmonary pathology. Thorax 17:320–328

20. Weibel ER, Gomez DM (1962) A principle for counting tissue structures on random sections. J Appl Physiol 17:343–348

21. Mazzolo A, Bt R, Gille W (2003) Properties of chord length distributions of nonconvex bodies. J Math Phys 44:6195

22. Santaló L (1976) Integral geometry and geometric probability. Addison-Wesley Publishing Company, Inc, London

23. Parameswaran H, Majumdar A, Ito S, Alencar AM, Suki B (2006) Quantitative characterization of airspace enlargement in emphysema. J Appl Physiol 100(1):186–193

24. Saetta M, Shiner RJ, Angus GE, Kim WD, Wang NS, King M, Ghezzo H, Cosio MG (1985) Destructive index: a measurement of lung parenchymal destruction in smokers. Am Rev Respir Dis 131:764–769

25. Thurlbeck WM (1967) Internal surface area and other measurements in emphysema. Thorax 22:483–496

26. Emami K, Cadman RV, Woodburn JM, Fischer MC, Kadlecek SJ, Zhu J, Pickup S, Guyer RA, Law M, Vahdat V, Friscia ME, Ishii M, Yu J, Gefter WB, Shrager JB, Rizi RR (2008) Early changes of lung function and structure in an elastase model of emphysema–a hyperpolarized 3He MRI study. J Appl Physiol 104:773–786. doi:10.1152/japplphysiol.00482.2007

27. Gillooly M, Lamb D (1993) Microscopic emphysema in relation to age and smoking habit. Thorax 48:491–495

28. Saetta M, Kim WD, Izquierdo JL, Ghezzo H, Cosio MG (1994) Extent of centrilobular and panacinar emphysema in smokers' lungs: pathological and mechanical implications. Eur Respir J 7:664–671

29. Ito S, Bartolak-Suki E, Shipley JM, Parameswaran H, Majumdar A, Suki B (2006) Early emphysema in the tight skin and pallid mice: roles of microfibril-associated glycoproteins, collagen, and mechanical forces. Am J Respir Cell Mol Biol 34(6):688–694

30. Suki B, Lutchen KR (1992) Pseudorandom signals to estimate apparent transfer and coherence functions of nonlinear systems: applications to respiratory mechanics. IEEE Trans Biomed Eng 39(11):1142–1151

31. Farre R, Rotger M, Navajas D (1992) Optimized estimation of respiratory impedance by signal averaging in the time domain. J Appl Physiol 73(3):1181–1189

32. Ito S, Ingenito EP, Arold SP, Parameswaran H, Tgavalekos NT, Lutchen KR, Suki B (2004) Tissue heterogeneity in the mouse lung: effects of elastase treatment. J Appl Physiol 97 (1):204–212

33. Ito S, Lutchen KR, Suki B (2007) Effects of heterogeneities on the partitioning of airway and tissue properties in normal mice. J Appl Physiol 102(3):859–869. doi:10.1152/ japplphysiol.00884.2006. 00884.2006 [pii]

Chapter 9

Practical Methods for Assessing Emphysema Severity Based on Estimation of Linear Mean Intercept (Lm) in the Context of Animal Models of Alpha-1 Antitrypsin Deficiency

Airiel M. Davis, Kristen E. Thane, and Andrew M. Hoffman

Abstract

Alpha-1 antitrypsin deficiency is typified by panacinar emphysema in humans. Whilst animal models of (α1A-TD) that more accurately reflect the histology and molecular pathology of α1A-TD are in development, it is timely to discuss methods to assess emphysema severity. Several methods exist to quantify emphysema from histologic sections, including linear mean intercept (Lm), equivalent diameters (D) or their statistical derivatives (D2), and more recently probability models of D2 ("severity index"). Given proper attention to lung inflation, reference volume, and random sampling, Lm determined by intersect point counting provides a robust analytical tool to quantify emphysema severity. Details of lung preparation, processing for random sampling, and batch processing of prescreened images are provided herein.

Key words Stereology, Lm, Lung morphometry, Lung surface area, Panacinar emphysema, Alpha-1 antitrypsin deficiency

1 Introduction

Alpha-1 antitrypsin deficiency (α1A-TD) causes severe panacinar emphysema in adult humans [1, 2]. Development of animal models for this heritable condition is critical to better understand the link between the heritable gene mutations, molecular pathology, and phenotype, and to advance novel therapies including gene correction for this disease. An important aspect of assessing phenotype of α1A-TD in animal models is quantification of emphysema severity. Linear mean intercept (Lm) [3, 4] is widely employed for this purpose [5–8]. The quantity Lm is an estimation of average free distance within the acinar structures including alveoli and alveolar ducts, which are expected to be uniformly dilated in α1A-TD. Lm is confounded by lung volume (due to recoil pressure or user defined inflation conditions) and by alveolar shape (reliant on isotropy), although precautions can be taken to avoid these pitfalls [9]. Several methods exist that improve sensitivity and accuracy of assessing

Florie Borel and Christian Mueller (eds.), *Alpha-1 Antitrypsin Deficiency: Methods and Protocols*, Methods in Molecular Biology, vol. 1639, DOI 10.1007/978-1-4939-7163-3_9, © Springer Science+Business Media LLC 2017

histologic severity of emphysema including the calculation of equivalent diameters (D) or their statistical derivatives (D2) [10, 11], and more recent probability models of D2 ("severity index") [12]. Analysis of animal models of α1A-TD may or may not derive benefit from these second-generation statistical models. The simplicity and large historical database of Lm in lung research are advantages to employing this index. Hence, basic methods to prepare the lung for estimation of Lm and lung surface area relevant to emphysema models are described in this section. More comprehensive discussions of Lm or stereology including three-dimensional analysis of alveolar numbers are available elsewhere [9, 13–17].

2 Materials

Lung preparation

1. Mouse dissection equipment (tools, board, pins).
2. 0.9% NaCl or 1× PBS, 10 mL for flushing, 10 mL syringe, 22 gauge needle.
3. 20 gauge 1 in. catheter.
4. 3-0 braided silk suture.
5. Ultra-low gelling temperature agarose (Sigma A2576), made up to 4% in 1× PBS.
6. 10% formalin, for agarose mixture and ~30 mL per mouse for lung fixation.
7. Heated water bath.
8. Medium weigh boat, one per mouse.
9. Metric ruler.
10. Ring stand assembly, stand, 30 mL syringe, two-way stopcock, flexible tubing, male Luer lock fitting.
11. Bucket with ice.

Total lung volume calculation

1. Dissection tools.
2. Glass beaker with water.
3. Balance.
4. Device to maintain lung under water.

Mean linear intercept calculation

1. Light microscope for bright field microscopy with 200× magnification.
2. Computer (PC)
3. Digital camera.

4. SigmaScan Pro5 software.

5. Macro.

6. Grid overlay.

2.1 Methods for Preparing Lung for Fixation

1. An agarose instillation system is built using a clamp stand, a 30 mL or larger syringe, a stopcock, a segment of flexible tubing (Tygon R-3603), and a male end standard Luer lock adaptor (Fig. 1).

2. The liquid agarose mixture is prepared by mixing equal volumes of liquefied 4% agarose in PBS (Sigma A2576) with prewarmed 10% neutral buffered formalin and kept at a temperature of 37–42 °C until ready to use (*see* **Note 1**).

3. Sacrifice the mouse by isoflurane inhalation overdose. Cervical dislocation may be used as a secondary euthanasia method.

4. Pin the mouse securely to a dissection board. Dissect the skin and subcutaneous tissue along the abdomen and thorax, and pin the skin away from the dissection field (Fig. 2).

5. Using sharp dissection, open the peritoneum and sever the caudal vena cava between the diaphragm and the liver (Fig. 3). Carefully pierce the diaphragm, taking care to avoid trauma to the pulmonary tissue. Sharply dissect away the

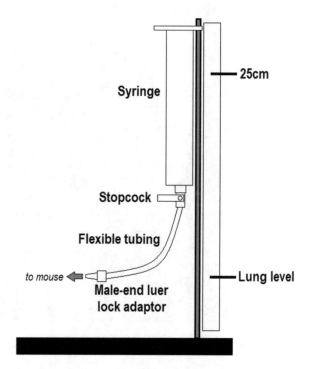

Fig. 1 Schematic of system for delivery of liquid agarose–formalin mixture for lung fixation

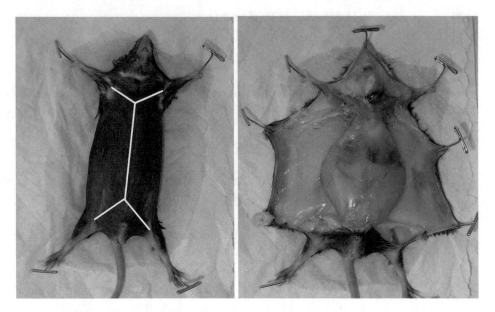

Fig. 2 Positioning for initial dissection of the mouse. Dissection is made through the skin and subcutaneous tissue as demonstrated by the white lines (*left*). The skin is pinned to stabilize further dissection (*right*)

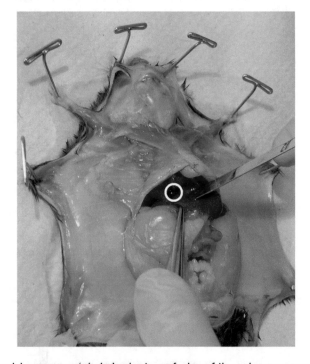

Fig. 3 Transect the caudal vena cava (*circled*) prior to perfusion of the pulmonary vasculature

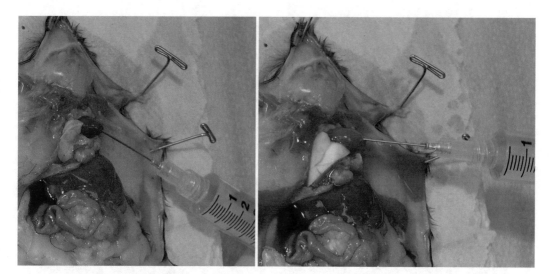

Fig. 4 Saline flush of the pulmonary vasculature via right ventricular puncture. As the flush progresses, the pulmonary tissues will appear white (*right*)

diaphragm. Remove the ribs and intercostal tissues as dorsally as possible to fully expose the lungs.

6. Flush the pulmonary vasculature with 10 mL of saline via puncture of the right ventricle (*see* **Note 2**), taking care to use gentle pressure to avoid overdistension of the lungs (Fig. 4).

7. Allow the lungs to deflate maximally to minimize the amount of fluid retention in the vasculature.

8. Make a careful incision through no more than 1/3 of the trachea immediately distal to the larynx. Remove a 20 gauge 1 in. intravenous catheter from its stylet (BD 381433). Gently insert the catheter into the trachea until minor resistance is felt. Do not force the catheter deeper, as this might lead to trauma of the airways or uneven filling distribution of the agarose mixture within the lungs. Stabilize the catheter within the trachea using fine gauge braided suture (4-0 silk or similar) passed circumferentially around the trachea, positioned a minimum of 1–2 tracheal rings below the level of catheter insertion to minimize slippage of the suture. A single square throw should be placed to stabilize the catheter in place, but a full knot should not be tied at this step (Fig. 5).

9. If desired, the lung may be degassed prior to instilling the agarose–formalin mixture. To perform degassing, the entire mouse (with chest cavity open and tracheal cannula in place) should be placed within a vacuum chamber. The vacuum should be applied for no longer than 3 min prior to proceeding to the next step.

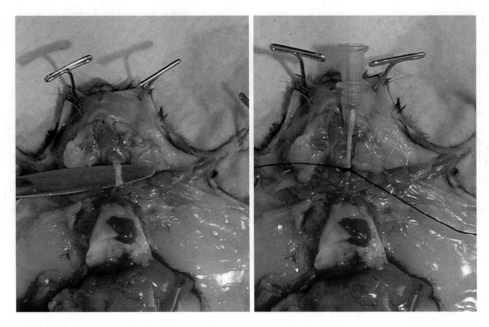

Fig. 5 Identification of the trachea (*left*) and intubation using a 20 gauge 1 in. catheter (*right*). The catheter is stabilized within the trachea with a single suture throw

10. The entire mouse is next placed within a medium weigh boat for optimal positioning of the lungs for filling with the 1:1 agarose–formalin mixture (Fig. 6). Prime the agarose instillation system (syringe and tubing) to remove all air from the system. The liquid agarose level should be filled to a position 25 cm above the level of the lungs to achieve correct filling pressure. Working rapidly, attach the free end of the tubing to the catheter secured within the trachea, taking care to not dislodge the catheter, and then allow the liquid agarose to flow into the lungs. Gentle pressure using the syringe plunger may be necessary to initiate flow into the lungs. The liquid agarose–formalin mixture can be refilled as needed to maintain a 25 cm filling pressure. Complete filling of the lungs is expected within 1–2 min, but may occur more rapidly (*see* **Note 3**). Once complete filling has been achieved, withdraw the catheter from the trachea and immediately tighten the suture placed around the trachea to prevent any backflow of the agarose mixture. A second suture throw may be placed to form a secure square knot.

11. Place the entire weigh boat on ice for 10 min to allow the agarose–formalin mixture to completely solidify. Sharply dissect the heart–lung bloc free, taking care to avoid trauma to the trachea or pulmonary tissues. Gentle traction can be applied to the trachea using the tied suture to facilitate dissection. Place the lung–heart bloc into 10% neutral buffered formalin for a minimum of 24 h prior to sectioning.

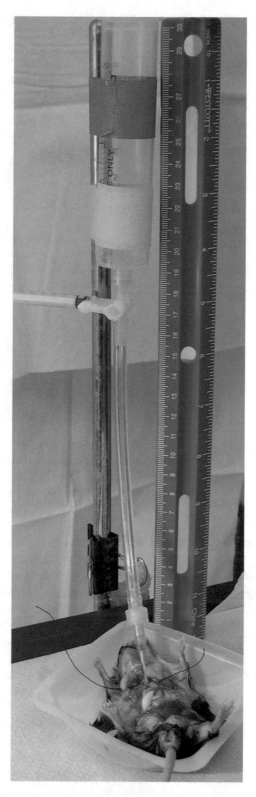

Fig. 6 Attaching the tracheal intubation catheter to the primed agarose instillation system. Ensure that the liquid agarose–formalin mixture fills the system to 25 cm above the level of the lungs to achieve proper filling pressure

Fig. 7 Total lung volume by water displacement. All nonpulmonary tissue must be removed following fixation. Correct placement of the lung must be achieved to ensure accuracy of measurement. The lung tissue must not touch the sides or bottom of the container and must be fully submerged below the surface of the water

2.2 Measurement of Total Lung Volume Using Volume Displacement

Total lung volume may be calculated prior to sectioning by utilizing water displacement technique. The heart and any other extraneous tissue must be removed (after fixation) prior to weighing to obtain an accurate calculation. The lungs must be positioned correctly to achieve an accurate measurement (*see* **Note 4**). The lung tissue cannot touch the bottom or sides of the container, and must be positioned below the surface of the water (Fig. 7).

2.3 Random Sampling

1. The fixed lung as a whole or individual lobes (tied off and excised) is/are randomly oriented in a cassette and embedded in paraffin, making sure the paraffin completely surrounds the lung tissue. Paraffin-embedded tissues are sectioned initially at a random depth (e.g., 25–75 μm) and thereafter regular intervals (e.g., every 50 μm) to obtain randomly sampled sections. The sections are stained with hematoxylin and eosin or toluidine.

2. Ten photomicrographs are taken using a 10 or 20× objective on a bright field microscope at locations selected randomly by way of a reticle and random number generator, excluding nonalveolar parenchymal structures but including alveoli and alveolar ducts.

3. Lm (chord length) is measured as the number of intercepts of acinar structures (alveolar, alveolar duct) to grid lines divided into the total length of the coherent grid line, excluding non-parenchymal structures and edge effects (long traverses along alveolar ducts, for instance). According to the original description of this method by Campbell and Tomkeieff [4] the surface/volume ratio is equivalent to four times the number of intercepts divided by the total length of the coherent line (sum of incoherent grid lines). Therefore, total alveolar surface area (units $= \mu m^2$) is computed using the following formula: SA $= [(4 \times \text{lung volume})/\text{mean Lm}] \times 1000$ [9], where lung volume is determined from volume displacement (described above). Estimation of Lm for a single field of view can be achieved by exporting each field of view image (e.g., as a .jpg) into digital scanning software in order to apply a grid (Fig. 8). A macro such as the one provided below for use in SigmaScan can be used to compute Lm on single images or batches placed in a single folder (*see* **Note 5**). For this macro, a digital grid is composed of 32 evenly spaced lines, each of 42 μm length for a total length of 1344 μm.

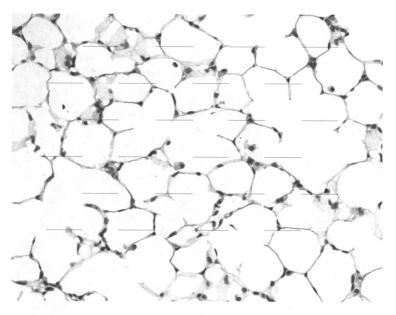

Fig. 8 Example of grid employed for intersect point counting system using SigmaScan in mouse lungs

Macro for batch processing of Lm from photomicrographs (.jpg):

```
Public App As Object
Public Worksheet As Object
Public ImageName As String
Public GridName As String
Public GridLength As Long
Public Image0 As Object
Public MLIoverlay0 As Object
Public Intercepts As Long
Public MLIntercept As Double

Sub Main
        'load SigmaScan functions
          Set App = CreateObject("SigmaScan.Application")

      'turn off all measurements
       For i=0 To NUMMEASURES-1
              App.DoNotCollectMeasurement(i)
              Next i

      GetInfo
      'Debug.Print "TextBox1:" + (ImageName) + "-"
      'Debug.Print "TextBox2:" + (GridName) + "-"
      'Debug.Print "TextBox3;" + CStr(GridLength) + "-"
      OpenImages
      Clearfield
              Threshold
              'RemoveObjects
              OverlayCalibGrid
              CountIntercepts
              MLI
              Name
              ResultCode = image0.HideOverlay(1)
              End

End Sub
'asks user for input
Sub GetInfo
        'get name and path of image
        MsgBox("Select image to be measured")
        ImageName = GetFilePath$()

        'get name and path of grid
        MsgBox("Select calibrated grid")
        GridName = GetFilePath$()

        'get total length of grid
        Begin Dialog UserDialog 400,203 ' %GRID:10,7,1,1
            Text 20,42,150,14,"Total length of grid:",.Text1
            TextBox 150,70,90,21,.TextBox1
            OKButton 150,126,90,21
            Text 260,77,90,14,"microns",.Text2
```

```
              End Dialog
              Dim dlg As UserDialog
              Dialog dlg
              GridLength = CLng(dlg.TextBox1)
    End Sub
    'specifies image to analyze and grid
    Sub OpenImages
              Set Image0 = App.OpenImage(ImageName)
              ResultCode = Image0.Show
        ResultCode = Image0.MakePermanent
              Set MLIoverlay0 = App.OpenImage(GridName)

    End Sub
    'perform pseudoclearfield equalization of image
    Sub Clearfield
          ResultCode = Image0.ChangeColorResolution(32, 4)
          ResultCode = Image0.ConvertToGrayScale
          Dim Image1 As Object
          Set Image1 = Image0.CreatePseudoClearField(5, 5)
          ResultCode = Image1.Show
          ResultCode = App.ClearFieldEqualization(Image0, Image1)

    End Sub
    'intensity threshold and erode image to reduce residues
    Sub Threshold
      ResultCode = Image0.ChangeColorResolution(8, 4)
      ResultCode = Image0.ConvertToGrayScale
      Dim Left0(1) As Long
      Left0(0) = 0
      Dim Right1(1) As Long
      Right1(0) = 225
      ResultCode = Image0.IntensityThreshold(1, 1, Left0, Right1)
      ResultCode = Image0.FilterOverlay(0, 1, 1, 2, 1)

    End Sub
    'remove small, round artifacts
    Sub RemoveObjects
              ' open Worksheet
              Set Worksheet = App.GetWorksheet
      Worksheet.Show
      Worksheet.MakePermanent
              '* Count the objects
              App.CollectMeasurement(35, "B")
      ResultCode = image0.CountObjects(1)
      NumItems = Worksheet.GetCellValue("B",1)

              '* For each object find its area and shape factor
              App.DoNotCollectMeasurement(35)
              App.CollectMeasurement(11, "A")
              App.CollectMeasurement(34, "C")
      ResultCode = image0.MeasureObjects(1)
```

```
                    '* Eliminate all objects not sufficiently round
                    MsgBox("Removing artifacts. Click OK to continue.")
                    iNum=0
             For ii = 1 To NumItems
                    Shape = Worksheet.GetCellValue("C",ii)
                    ObjArea = Worksheet.GetCellValue("A",ii)
                    If (Shape > 0.5) And (ObjArea < 10) Then
                            '* Eliminate the object
                            ResultCode = image0.EliminateObject(ii)
                            iNum=iNum+1
                    End If
             Next ii
             MsgBox(iNum+" artifacts removed. Click OK to finish.")
      End Sub
      ' superimpose calibrated grid on image
      Sub OverlayCalibGrid
             ResultCode = MLIoverlay0.Copy(0, 0, 1187, 970, True, 3)
          ResultCode = Image0.Paste(100, 130, True, 3)'location of grid on
      image
             ResultCode = Image0.AndOverlays(1, 3, 4)
             ResultCode = Image0.FilterOverlay(4, 4, 4, 10, 5)

      End Sub
      ' count intercepts
      Sub CountIntercepts
             'App.CollectMeasurement(M_NUMOBJECTS, "B")
             Intercepts = Image0.CountObjects(4)
             'Result = MsgBox("Intercepts:" + CStr(Intercepts))
             'Set Worksheet = App.OpenWorksheet("C:\Documents and
      Settings\ltsai\My Documents\MLI Template2.xls")

      End Sub
      ' calculate Mean Linear Intercept
      Sub MLI
             MLIntercept = GridLength / Intercepts
             MsgBox("Mean Linear Intercept:" + CStr(MLIntercept))

      End Sub
      ' add name to worksheet

      Sub Name
             Set Worksheet = App.GetWorksheet
             ResultCode=Worksheet.SetCellText("A", 1, CStr(ImageName))
             ResultCode = Worksheet.SetCellValue("A", 2, MLIntercept)
             Worksheet.Show
             Worksheet.MakePermanent
      End Sub
      Sub ListFiles(strPath As String, Optional Extention As String)
      'Leave Extention blank for all files
      Dim File As String
      If Right$(strPath, 1) <> "\" Then strPath = strPath & "\"
             If Trim$(Extention) = "" Then
                    Extention = "*.*"
```

```
              ElseIf Left$(Extention, 2) <> "*." Then
                    Extention = "*." & Extention
        End If
        File = Dir$(strPath & Extention)
        Do While Len(File)
                List1.AddItem File
          File = Dir$
        Loop

        End Sub
```

---

3 Notes

1. The agarose–formalin mixture should never be instilled into the lung at temperatures greater than 42 °C as this can lead to architectural artifacts.

2. Mice may be given heparin prior to sacrifice to prevent coagulation and to aid in flushing the vasculature. Throughout the procedure, it is imperative to work rapidly to minimize complications from coagulation within the pulmonary vasculature and avoid uneven alveolar filling due to premature cooling and solidification of the agarose.

3. If the agarose mixture gels too quickly and complete filling is not achieved, keeping the mouse warm or working more quickly may be required. Ensure that all equipment and supplies are prepared and the agarose instillation system is set up prior to beginning the procedure.

4. When performing lung volume assessment using water displacement, it is necessary for the lung to be positioned correctly within the water container on the scale. If needed, a device can be fashioned from a piece of stainless steel wire shaped into a small spiral-like basket to submerge the lungs under the water surface and prevent the tissue from coming into contact with the sides of the container.

5. Prior to initiating computed calculation of Lm using SigmaScan, it is important to discard photomicrographs that contain nonlung artifacts (densities) or nonparenchymal structures (blood vessels, inflammatory cells, red blood cells, and bronchiolar lumens) that might be identified as point intersections, as this will result in underestimation of Lm. For the purposes of this description, 'automation' is limited to batch processing of prescreened high quality images.

Acknowledgments

The authors thank Aaron M. Burgess for generating the code used to automate analysis of Lm from image batches. Funding from the Shipley Foundation (AMH) and NHLBI: HL112987-01A1 (AMH/Edward Ingenito).

References

1. Zuo L, Pannell BK, Zhou T et al (2016) Historical role of alpha-1-antitrypsin deficiency in respiratory and hepatic complications. Gene. doi:10.1016/j.gene.2016.01.004

2. Greulich T, Vogelmeier CF (2016) Alpha-1-antitrypsin deficiency: increasing awareness and improving diagnosis. Ther Adv Respir Dis 10(1):72–84. doi:10.1177/1753465815602162

3. Dunnill MS (1964) Evaluation of a simple method of sampling the lung for quantitative histological analysis. Thorax 19:443–448

4. Campbell H, Tomkeieff SI (1952) Calculation of the internal surface of a lung. Nature 170 (4316):116–117

5. Andersen MP, Parham AR, Waldrep JC et al (2012) Alveolar fractal box dimension inversely correlates with mean linear intercept in mice with elastase-induced emphysema. Int J Chron Obstruct Pulmon Dis 7:235–243. doi:10.2147/COPD.S26493

6. Hoffman AM, Shifren A, Mazan MR et al (2010) Matrix modulation of compensatory lung regrowth and progenitor cell proliferation in mice. Am J Physiol Lung Cell Mol Physiol 298(2):L158–L168. doi:10.1152/ajplung.90594.2008

7. Lee J, Reddy R, Barsky L et al (2009) Lung alveolar integrity is compromised by telomere shortening in telomerase-null mice. Am J Physiol Lung Cell Mol Physiol 296(1):L57–L70. doi:10.1152/ajplung.90411.2008

8. Choe KH, Taraseviciene-Stewart L, Scerbavicius R et al (2003) Methylprednisolone causes matrix metalloproteinase-dependent emphysema in adult rats. Am J Respir Crit Care Med 167(11):1516–1521

9. Knudsen L, Weibel ER, Gundersen HJ et al (2010) Assessment of air space size characteristics by intercept (chord) measurement: an accurate and efficient stereological approach. J Appl Physiol (1985) 108(2):412–421. doi:10.1152/japplphysiol.01100.2009

10. Parameswaran H, Majumdar A, Ito S et al (2006) Quantitative characterization of airspace enlargement in emphysema. J Appl Physiol 100(1):186–193

11. Wilson AA, Murphy GJ, Hamakawa H et al (2010) Amelioration of emphysema in mice through lentiviral transduction of long-lived pulmonary alveolar macrophages. J Clin Invest 120(1):379–389. doi:10.1172/JCI36666

12. Marcos JV, Munoz-Barrutia A, Ortiz-de-Solorzano C et al (2015) Quantitative assessment of emphysema severity in histological lung analysis. Ann Biomed Eng 43(10):2515–2529. doi:10.1007/s10439-015-1251-5

13. Schneider JP, Ochs M (2013) Stereology of the lung. Methods Cell Biol 113:257–294. doi:10.1016/B978-0-12-407239-8.00012-4

14. Hsia CC, Hyde DM, Ochs M et al (2010) An official research policy statement of the American Thoracic Society/European Respiratory Society: standards for quantitative assessment of lung structure. Am J Respir Crit Care Med 181(4):394–418. doi:10.1164/rccm.200809-1522ST

15. Ochs M (2014) Estimating structural alterations in animal models of lung emphysema. Is there a gold standard? Ann Anat 196 (1):26–33. doi:10.1016/j.aanat.2013.10.004

16. Hsia CC, Hyde DM, Ochs M et al (2010) How to measure lung structure – what for? On the "Standards for the quantitative assessment of lung structure". Respir Physiol Neurobiol 171 (2):72–74. doi:10.1016/j.resp.2010.02.016

17. Weibel ER, Hsia CC, Ochs M (2007) How much is there really? Why stereology is essential in lung morphometry. J Appl Physiol 102 (1):459–467. doi:10.1152/japplphysiol.00808.2006

Chapter 10

Design, Cloning, and In Vitro Screening of Artificial miRNAs to Silence Alpha-1 Antitrypsin

Florie Borel and Christian Mueller

Abstract

This protocol describes the design, cloning, and in vitro screening of artificial microRNAs (miRNAs) to silence alpha-1 antitrypsin (AAT). This method would be of interest to silence AAT in a variety of in vitro or in vivo models, and prevalidated sequences against human AAT are provided. This simple 5-day protocol may more generally be used to design artificial miRNAs against any transcript.

Key words Alpha-1 antitrypsin, AAT, Z-AAT, RNA interference, RNAi, microRNA, miRNA, Short hairpin RNA, shRNA

1 Introduction

RNA interference (RNAi) is a powerful molecular tool to achieve silencing of a target messenger RNA (mRNA) of interest, here, alpha-1 antitrypsin (AAT). Artificial microRNAs (miRNAs) are one type of RNAi effectors that represent an easy-to-design strategy to achieve silencing that can be quickly implemented in any molecular biology lab. In this chapter, we describe how to design, clone, and screen in vitro artificial miRNAs targeting AAT. Artificial miRNAs present the advantage that they can be expressed from a polymerase II (pol II) or polymerase II (pol III) promoter, and they can easily and safely be used in vivo. Other types of RNAi effectors (in particular short hairpin RNAs and small interfering RNAs) are not described here.

Florie Borel and Christian Mueller (eds.), *Alpha-1 Antitrypsin Deficiency: Methods and Protocols*, Methods in Molecular Biology, vol. 1639, DOI 10.1007/978-1-4939-7163-3_10, © Springer Science+Business Media LLC 2017

2 Materials

1. miRNA guide sequence of interest as identified in Subheading 3.

2. Expression plasmid in which to clone the sequence of interest, containing either a polymerase II or polymerase III promoter (H1, U6) with or without a reporter gene such as GFP.

3. Cloning software (optional).

4. Nuclease-free duplex buffer (Integrated DNA Technologies, Coralville, IA).

5. Qiagen plasmid plus midi kit (Cat#12943, Qiagen, Germantown, MD).

6. Restriction enzymes of choice (New England Biolabs, Ipswich, MA or Thermo Scientific, Waltham, MA).

7. Electrophoresis-grade agarose.

8. Gel electrophoresis equipment.

9. Clean razor blades.

10. QIAquick gel extraction kit (Cat# 28704, Qiagen).

11. Quick ligation kit (Cat# M2200S, New England Biolabs).

12. SURE 2 supercompetent cells (Cat# 200152, Agilent Technologies, Santa Clara, CA).

13. NZY$^+$ broth (Cat# N1225, Teknova, Hollister, CA).

14. Selection plates with antibiotic of choice (depending on selected expression plasmid) for the selection of positive clones.

15. Terrific broth with antibiotic of choice (depending on selected expression plasmid) for growth of positive clones.

16. Cell line of specie of choice expressing alpha-1 antitrypsin, which can be transfected.

17. Transfection method of choice (depending on selected cell line).

18. Trizol reagent (Cat#15596–018, Life Technologies, Carlsbad, CA).

19. Nuclease-free water (Cat#AM9937, Life Technologies).

20. Nanodrop (Nanodrop, Wilmington, DE).

21. High-capacity RNA-to-cDNA (Cat# 4387406, Life Technologies).

22. PCR tubes.

23. PCR machine.

24. Taqman primers and probe.

25. Taqman fast advanced mastermix (Cat# 4444557, Life Technologies).

26. RT-qPCR plates.

27. RT-qPCR machine.

3 Methods

DAY 1

1. Retrieve the mRNA sequence for alpha-1 antitrypsin of your specie of interest (or consider identifying conserved stretches of sequence, *see* **Note 1**) from NCBI.

2. Fold the entire mRNA in a RNA folding program (such as RNAfold) and determine stretches of sequence that are predicted to be accessible (*see* **Note 2**).

3. In those, select 21-nt long miRNA guide sequence(s) of interest (*see* **Note 3**) starting with A or U. Alternatively, one of the sequences targeting human AAT previously validated by our lab [1] may be used (Table 1 and Fig. 1, available on request at www.umassmed.edu/muellerlab). Further refine your selection by analyzing the predicted off-target profile of your sequence by performing a nucleotide BLAST search against the genome of the specie(s) that the construct will be used in.

4. Design the artificial miRNA: incorporate the miRNA guide sequence in the miRNA backbone of choice (*see* **Note 4** and Table 2), as described in Table 3 and Fig. 1. Incorporate to the design the restriction site(s) of choice (depending on selected expression vector) at each end (*see* **Note 5**), so that cohesive ends will be obtained when annealing the oligos. Have both forward and reverse strands oligos synthesized.

DAY 2

5. Anneal oligos: mix 1 μl of each oligo (forward and reverse) at 100 μM in 8 μl of annealing buffer (Nuclease-free duplex buffer, IDT). Heat to 95 °C for 5 min and then cool down slowly (*see* **Note 6**).

Table 1
List of validated artificial miRNA sequences targeting human AAT

miR-914 guide	5′-AATGTAAGCTGGCAGACCTTC-3′
miR-910 guide	5′-TAAGCTGGCAGACCTTCTGTC-3′
miR-943 guide	5′-ATAGGTTCCAGTAATGGACAG-3′

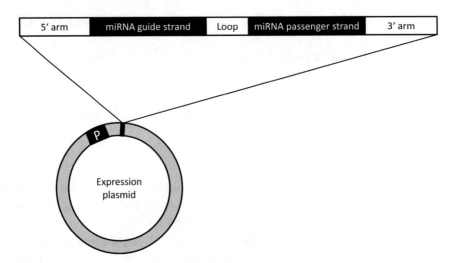

Fig. 1 Design of the artificial miRNA construct. The artificial miRNA will be composed of a 5′ arm, a miRNA guide strand, a loop, a miRNA passenger strand, and a 3′ arm. The 5′ arm, the loop, and the 3′ arm, represented in white, are predesigned sequences that originate from cellular miR-155. The miRNA guide and passenger strands, represented in *black*, are custom designs. *P* promoter

Table 2
Sequence of miR-155 backbone used in this protocol

5′ arm backbone	5′-CCTGGAGGCTTGCTGAAGGCTGTATGCTG-3′
Loop backbone	5′-GTTTTGGCCACTGACTGAC-3′
3′ arm backbone	5′-CAGGACACAAGGCCTGTTACTAGCACTCACATGGAACAAATGGCC-3′

Table 3
Sequence assembly to build an artificial miRNA based on the miR-155 backbone

5′ arm backbone	5′-CCTGGAGGCTTGCTGAAGGCTGTATGCTG-3'
miR-914 guide	5′-AATGTAAGCTGGCAGACCTTC-3'
Loop backbone	5′-GTTTTGGCCACTGACTGAC-3'
miR-914 passenger	5′-GAAGGTCT_ _CAGCTTACATT-3'
3′ arm backbone	5′-CAGGACACAAGGCCTGTTACTAGCACTCACATGGAACAAATGGCC-3'

The underscore characters in the passenger strand sequence mark the bases that are absent as compared to the guide strand. The two bases missing in the middle of the passenger strand will create a bulge

6. Prepare the expression plasmid DNA (Qiagen plasmid plus midi kit). Digest at least 1 μg of the expression plasmid with the selected restriction enzyme(s) for 2 h at 37 °C (*see* **Note** 7).

7. Run 2 μl of the annealed oligos and 2 μl of the digested expression plasmid on a 1% agarose gel with molecular ladders of the appropriate size (1 kb and 100 bp sizes are suggested for

the plasmid and oligos respectively) to verify proper annealing of the oligos and proper digestion of the expression plasmid (*see* **Note 8**).

8. Cut the band of interest from the expression plasmid digestion using a clean razor blade. Proceed to gel extraction (QIAquick gel extraction kit, Qiagen) according to the manufacturer's recommendations (*see* **Note 9**).

9. Ligate (Quick ligation kit, NEB) the annealed oligos into the expression plasmid according to the manufacturer's recommendations (*see* **Note 10**). Briefly:

   - 10 μl 2× quick ligase buffer.
   - 1 μl quick ligase.
   - 50 ng digested, gel-purified expression plasmid (estimate concentration based on gel).
   - × μl annealed oligos (3:1 oligos: plasmid ratio).
   - Up to 20 μl dH$_2$O.

   Mix, spin down, and incubate for 5 min at RT.

10. Transform in SURE 2 supercompetent cells according to the manufacturer's recommendations (*see* **Note 11**). Briefly:

    - 100 μl SURE 2 supercompetent cells.
    - 2 μl 1.22 M beta-mercaptoethanol.
    - 1 μl ligation.
    - 0.9 ml pre-heated NZY$^+$ broth.

DAY 3

11. The next day, inoculate minipreps and grow at 30 °C, 250 rpm (*see* **Note 12**).

12. After 8 h of culture, isolate plasmid DNA. Check for the presence of the miRNA insert in the plasmid using the restriction enzymes used for cloning and running the digestion on a 1% agarose gel (*see* **Note 13**).

13. Send the positive clone(s) for sequencing to confirm the absence of mutation in the miRNA sequence (*see* **Note 14**) and prepare transfection-grade DNA (Qiagen plasmid plus midi kit).

DAY 4

14. Transfect the cell line of choice (*see* **Note 15**). Seed cells in 6-well plates so that they reach 70–90% confluence at the time of transfection. Transfect 2.5 μg of plasmid DNA per well using the transfection method of choice according to the manufacturer's recommendations. Include a mock transfection control and a miRNA control, *i.e.*, an artificial miRNA designed to target a mRNA other than AAT.

DAY 5

15. At 48 h post-transfection, harvest the cells in Trizol (Life Technologies) and isolate RNA according to the manufacturer's recommendations. Resuspend RNA in nuclease-free water (Life Technologies). Quantify RNA using a Nanodrop and homogenize all concentrations to normalize for retro-transcription efficiency.

16. Retro-transcribe RNA samples using the High Capacity RNA-to-cDNA kit (Life Technologies), according to the manufacturer's recommendations. Briefly:
    - 10 µl 2× RT buffer mix.
    - 1 µl 20× RT enzyme mix.
    - 9 µl RNA 200 ng/µl.
      Incubate for 60 min at 37 °C, 5 min at 95 °C, ∞ at 4 °C.

17. Quantify AAT mRNA and a housekeeping mRNA by RT-qPCR, according to the manufacturer's recommendations. Briefly:
    - 5 µl 2× Taqman fast advanced mastermix (Life Technologies).
    - 0.5 µl 20x Taqman probe of choice.
    - 4.5 µl cDNA 30 ng/µl (3× dilution).

18. Select the artificial miRNA that leads to the highest knockdown.

4   Notes

1. Instead of selecting a single transcript from one particular specie, one may select several transcripts from relevant species, perform a sequence alignment, and select conserved sequences for targeting. The main advantage of this strategy is that it allows validation of the same molecule in all preclinical studies from rodents to nonhuman primates, and can ultimately be taken to clinical testing.

2. For robustness, it is recommended to do the RNA folding predictions several times using different algorithms and select stretches of sequence that are consistently predicted to be accessible.

3. We recommend selecting 5–10 sequences to maximize the chances to get at least one construct that will elicit a high knockdown in vitro.

4. We recommend using miR-155 as a backbone, which is the sequence used in this protocol. However, other miRNA backbones may be used.

5. The restriction sites are chosen based on which cloning sites will be used to clone the artificial miRNA insert into the expression plasmid.

6. The slow cooling is critical for proper annealing. This can be done either in a PCR machine (-1 °C/min steps) or manually in a water bath (let the water bath cool down at RT).

7. Temperature may need to be adjusted depending on the selected restriction enzyme.

8. This validation step is critical and should be carefully performed prior to proceeding with the cloning. The annealed oligos should appear as one band at the expected size. If it is not the case, anneal the oligos again.

9. For a higher purity and higher subsequent ligation efficiency, we recommend following the gel extraction with a PCR purification step (QIAquick PCR Purification Kit, Cat#28104, Qiagen). To increase DNA recovery and concentration, elution may be used in 30 μl elution buffer pre-warmed to 50 °C.

10. It is recommended to include a negative control where the insert is replaced by dH_2O.

11. The use of SURE 2 supercompetent cells is recommended to minimize rearrangement and deletion of secondary structures.

12. A lower culture temperature of 30 °C (instead of the usual 37 °C) is recommended in order to minimize rearrangement of the plasmid that could be caused by the artificial miRNA secondary structure.

13. To optimize the timing of this protocol, we recommend doing a 5 min digestion using Fast Digest (Thermo Scientific) or Time-Saver (NEB) restriction enzymes.

14. Due to the presence of secondary structure, proper reads may be difficult to obtain. We recommend ordering "power reads" or consulting with your sequencing service.

15. For an efficient transfection, we suggest using HEK293T cells when targeting human AAT, and Hepa1–6 for murine AAT. If the expression plasmid does not contain a reporter gene, it is recommended to include a GFP or other reporter plasmid of your choice in an additional well to assess transfection efficiency. Determining knockdown may be challenging in difficult-to-transfect cells lines. For those, we recommend including a GFP in the vector, sorting cells, and determining knockdown in GFP-positive cells compared to GFP-negative cells. Finally, when targeting species for which a cell line suitable for screening is not available, HEK293T cells can be co-transfected with a plasmid expressing the target mRNA.

References

1. Mueller C, Tang Q, Gruntman A, Blomenkamp K, Teckman J, Song L, Zamore PD, Flotte TR (2012) Sustained miRNA-mediated knockdown of mutant AAT with simultaneous augmentation of wild-type AAT has minimal effect on global liver miRNA profiles. Mol Ther 20 (3):590–600. doi:10.1038/mt.2011.292. mt2011292 [pii]

Chapter 11

Methods to Identify and Characterize siRNAs Targeting Serpin A1 In Vitro and In Vivo Using RNA Interference

Stuart Milstein, Kun Qian, Linying Zhang, and Alfica Sehgal

Abstract

RNAi is a powerful tool that can be used to probe gene function as well as for therapeutic intervention. We describe a workflow and methods to identify screen and select potent and specific siRNAs in vitro and in vivo using qPCR-based methods as well as an AAT activity assay. We apply these techniques to a set of siRNAs targeting rat AAT, and use this set to exemplify the cell-based and in vivo data that can be generated using these methods.

Key words RNAi, siRNA, qPCR, AAT, Z-AAT, PiZ

1 Introduction

1.1 Alpha-1 Antitrypsin and Disease

Alpha-1 antitrypsin (AAT) is a serine protease inhibitor, synthesized by the liver and efficiently secreted into serum. One of its primary roles in the body is to inactivate neutrophil elastase in the lung and protect parenchymal tissue from nonspecific injury during periods of inflammation [1].

Alpha-1 antitrypsin deficiency (AATD) is caused by mutations in the gene that encodes AAT, *SERPINA1* [2, 3]. Although the *SERPINA1* gene is highly polymorphic with more than 90 variant alleles identified [4], most individuals (>95%) with manifest AATD are homozygous for the Glu342Lys mutation, encoded by the Z variant of AAT [5]. Individuals homozygous for the Z allele are said to have inherited the PiZZ protein phenotype. In PiZZ patients, the Z variant AAT protein (Z-AAT) folds abnormally during synthesis in the endoplasmic reticulum of hepatocytes and is retained intracellularly, rather than efficiently secreted. This misfolding of Z-AAT manifests as two distinct disease phenotypes, namely:

1. Loss of function: The lack of circulating anti-protease activity leaves the lung vulnerable to injury and the development of emphysema.

Florie Borel and Christian Mueller (eds.), *Alpha-1 Antitrypsin Deficiency: Methods and Protocols*, Methods in Molecular Biology, vol. 1639, DOI 10.1007/978-1-4939-7163-3_11, © Springer Science+Business Media LLC 2017

2. Toxic gain of function: The excessive intracellular accumulation of Z-AAT protein within hepatocytes leads to fibroinflammatory liver injury, cirrhosis, and hepatocellular carcinoma.

Although the molecular mechanisms by which accumulated polymerized Z-AAT protein cause liver pathology are not completely understood, evidence suggests that individuals with AATD who develop liver disease exhibit a lag in the protein degradation pathways critical for clearance of AAT polymers [6]. Support for this hypothesis comes from studies that show Z-AAT protein degradation is slower in cells of patients with liver disease compared to those without [7]. Accumulated AAT protein in the ER can trigger autophagy, mitochondrial injury, and apoptosis. Over time, the constitutive activation of stress and inflammatory pathways triggered by the presence of excess polymerized AAT protein in the liver leads to fibrosis, which can progress to cirrhosis and hepatocellular carcinoma (HCC).

1.2 RNA Interference

RNA interference, or RNAi, is a naturally occurring cellular mechanism for regulating gene expression that is mediated by "small interfering RNAs" (siRNAs). Typically, synthetic siRNAs are 19–25 base pair double-stranded oligonucleotides in a staggered duplex with a 2–4 nucleotide overhang at one or both of the 3′ ends. Such siRNAs can be designed to target the messenger RNA (mRNA) transcript of a given gene. When introduced into cells, the guide (or antisense) strand of the siRNA loads into an enzyme complex called the RNA-Induced Silencing Complex (RISC). This enzyme complex subsequently binds to its complementary mRNA sequence, mediating cleavage of the target mRNA and the suppression of the target protein encoded by the mRNA [8]. The ability to selectively degrade the mRNA encoding Z-AAT protein offers a potentially attractive approach for the treatment of liver disease associated with alpha-1 antitrypsin deficiency, and siRNAs are being evaluated in clinical trials.

Since unmodified siRNAs are rapidly eliminated and do not achieve significant tissue distribution upon systemic administration [9], various formulations are currently used to target their distribution to tissues, as well as to facilitate uptake of siRNAs into the relevant cell type.

This chapter describes one such approach that involves delivery of conjugated siRNA through the asialoglycoprotein receptor (ASGPR). The ASGPR is a member of the C-type lectin family of receptors that recognizes and binds glycoproteins with terminal, N-acetylgalactosamine (GalNAc) residues [10]. It is expressed on the cell surface of hepatocytes at a high copy number (0.5–1 million per cell) [11, 12], and facilitates clearance of desialylated glycoproteins from the blood [13]. Binding of the carbohydrate ligand to ASGPR leads to receptor-mediated endocytosis of the ligand-receptor complex, followed by the release of its cargo in the endocytic pathway, and subsequent recycling of the receptor back to the cell surface for successive rounds of uptake. Due to its high level and specificity of expression on hepatocytes, ability to mediate multiple rounds of uptake, and its ligand specificity, the ASGPR has been used in animals for liver-specific drug and gene delivery [14]. Multivalent N-GalNAc-conjugated siRNA have been shown to distribute to liver and have hepatocyte specific target silencing [15]. The methods described in this chapter are compatible with GalNAc-conjugated siRNAs.

1.3 siRNA Design

When designing siRNAs to target AAT or other therapeutic targets, three key attributes must be considered; a) siRNA potency, which is defined as the siRNA concentration required for half maximal silencing of the target transcript or IC_{50}; b) cross-reactivity to the same target in other species, if desirable (e.g., mouse, rats, nonhuman primates, or humans), and c) specificity, the ability to target only the intended target to the exclusion of the rest of the transcriptome [16]. While this chapter will focus mainly on in vitro and in vivo analysis of siRNAs targeting AAT, we will briefly cover these topics as well.

Although one approach to identify the most efficacious siRNA is by systematically walking siRNAs across the target sequences, it is more desirable to predict which siRNAs are more likely to work and screen only that subset. Alnylam uses a proprietary algorithm to predict efficacy which has been trained on a large set of activity measures generated against dozens of different targets. This algorithm is of limited utility to the public, since it was trained on siRNAs that contain specific patterns of chemical modification, but many other prediction and design algorithms are available in the public domain. While many features may differ between different tools, most are driven by similar variables including GC content and thermodynamics of asymmetric duplex unwinding. Such tools are often available on siRNA vendor websites.

1.4 siRNA Cross-Reactivity and Specificity

Another consideration in siRNA design is species cross-reactivity. It can be advantageous to have therapeutic siRNAs that are active across multiple species, as that allows exploring on-target and off-target effects with the same molecule, without having to use a surrogate molecule. While advantageous, a search for cross-reactive molecule usually restricts the pool of siRNAs with good predicted efficacy.

siRNA specificity, the ability of a siRNA to silence only the intended target, is an important feature that must be considered in siRNA design. siRNAs that are able to silence other targets, in addition to their intended target, may result in spurious results that may confound interpretation of the phenotypic consequences of on-target silencing. siRNAs are designed such that they have the maximum number of mismatches to all other transcripts, especially in the seed region (2–9 base of antisense).

1.5 Workflow   When identifying siRNAs to use as potential therapeutics or tool molecules to study a disease such as Alpha-1 antitrypsin deficiency, the methods below are followed. In a typical workflow, a set of siRNAs are designed and synthesized based on predicted efficacy, cross-species cross-reactivity, and specificity. TaqMan probes are validated to ensure that a single amplicon of the expected size is generated and that the amplification efficiency is optimal. This is typically done by performing qPCR on a serially diluted cDNA generated from the same cell line that will be used for the siRNA evaluation and by ensuring that the amplification occurs at 100% efficiency over the range of cDNA concentrations that will define the linear range of the assay (i.e., an increase of 1 Cp value for each doubling of input cDNA). Following TaqMan assay validation, siRNAs are tested in single or two dose screens to identify the most active siRNAs and eliminate those that are inactive. AAT targeting siRNAs are evaluated relative to naive or nontargeting siRNAs transfected under the same conditions. Ideally, siRNAs are tested in replicate experiments from independent starting cell cultures. The most active siRNAs are then tested in dose response screens to identify the most potent duplexes (those with the lowest IC_{50} values). Finally, a set of potent siRNAs are tested in vivo to select the molecule with maximum potency and durability (Fig. 1).

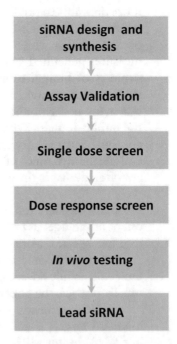

Fig. 1 siRNA selection workflow to identify potent siRNAs. The workflow to identify siRNAs consists of siRNAs design, synthesis, and TaqMan probe validation. This is followed by screening siRNAs in single-dose screens and a dose response. Most potent siRNA are then tested in vivo to identify lead siRNA molecule

2   Materials

1. Trasfection: Media as appropriate for cell type- for Hep3B EMEM supplemented with 10% FBS (ATCC), RNAiMax (ThermoFisher, cat no.: 13778-150), Trypsin 0.25%, 96-well tissue culture plates.

2. RNA purification: MagMAX 96 Total RNA isolation kit (ThermoFisher, cat no. AM1830), 96-well round-bottom plates (VWR, cat no. 29442-392).

3. cDNA synthesis: High-Capacity cDNA Reverse Transcription Kit with RNase Inhibitor (ThermoFisher cat no. 4374967), 96-well skirted PCR plates.

4. qPCR: LightCycler 480 Probes 2× Master Mix (Roche cat no. 04902343001) or as appropriate for real-time qPCR platform. LightCycler 480 Multiwell Plates.

5. ThermoFisher TaqMan Assays: Rat GAPDH: cat no. 4352338E, Human GAPDH: cat no. 4326317E, Rat AAT: Assay ID Rn00574670_m1, Human AAT: Assay ID Hs00165475_m1.

6. Protein measurements: ELISA kit to detect human AAT Abcam:ab108799. AAT Activity assay: Neutrophil Elastase Inhibitor Screening Kit (Abcam cat no. ab118971).

3   Methods

3.1 Transfection of Mammalian Cells with siRNAs Targeting AAT

siRNAs are introduced by reverse transfection, which should be carried out in a laminar flow hood following aseptic procedures at room temperature, unless otherwise specified.

1. Dilute siRNA to an appropriate concentration. For example, for 10 nM screen dilute each siRNA to 200 nM in 1× PBS such that 5 µl can be added for a 100 µl transfection reaction to give a final siRNA concentration of 10 nM. Dilute accordingly for other final siRNA concentrations.

2. Add 5 µl of each diluted siRNA to a well in a 96-well tissue culture plate and distribute a mixture containing 0.2 µl of RNAiMax in 14.8 µl of Opti-MEM to each well, mix by pipetting up and down, and incubate for 20 min.

3. While the mixture incubates, aspirate media from freshly split cells that have been grown to 80–90% confluence. Rinse cells with 5 ml of 0.25% trypsin.

4. Aspirate and replace with 3 ml of 0.25% trypsin for a 75 cm^2 flask and incubate at 37 °C for approximately 5 min or until cells have detached from the surface.

5. Add 27 ml of fresh media (without antibiotics if they are routinely used in your tissue culture practice) to inactivate the trypsin and transfer to 50 ml conical tube and centrifuge at $1500 \times g$ for 3 min.

6. Remove media from the 50 ml conical tube being careful not to disrupt the cell pellet and resuspend cells with 5–10 ml of complete media without antibiotics.

7. Count cells using a hemocytometer or automated counter such as a vi-cell to determine the total volume of cells that will be needed to transfect each well. We use 2×10^4 cells per well or 2×10^6 cells per 96-well plate. Adjust the volume to 2.5×10^5 cells/ml using complete media so that 80 µl of contains 2×10^4 cells.

8. Transfer 80 µl of cell suspension to each well of the 96-well plate and incubate for 24 h at the appropriate temperature and CO_2 concentration.

3.2 Isolation of RNA from Transfected Cells Using MagMAX Magnetic Bead Purification

RNA can be isolated from the wells using any method of choice. The protocol described here uses MagMAX magnetic beads, which allows for high-throughput processing of multiple plates. This method can also be adapted to 96-well liquid handlers if assays will be performed in high throughput.

1. Prepare reagents according to directions in the MagMAX-96 Total RNA Isolation Kit.

2. Vortex beads for 15 s, add 1.1 ml of beads to 1.1 ml of Lysis/Binding Enhancer for each 96-well plate to be isolated, and distribute 20 µl of the mixture to each well of a 96-well round-bottom plate.

3. Add 140 µl of Lysis/Binding buffer to previously transfected cells and shake at 650 rpm in a plate shaker.

4. Add lysed cells to magnetic bead mixture and shake for 5 min at 650 rpm and then place the round-bottom plate on a magnetic bead stand for 1 min.

5. For this and all subsequent washes, remove solution by aspiration or by carefully inverting the round-bottom plate on the magnetic ring stand.

6. Wash beads with 150 µl of Wash Solution 1, shake for 1 min at 650 rpm on a plate shaker. Place the plate on a magnetic stand for 1 min and then remove the solution. Repeat with Wash Solution 2.

7. Add 50 µl of DNase solution to each well and shake at 650 rpm for 15 min and then add 100 µl of RNA Rebinding Solution and shake for 3 min at 650 rpm. It is important to note that the RNA is liberated from the beads during the DNase treatment and will only be rebound upon the addition of the Rebinding Solution. Place on a magnetic stand for 1 min and aspirate.

8. Perform a final wash with 150 μl of Wash Solution 2 as in **step 6** this time making sure to remove as much of the residual wash buffer as possible and dry beads by shaking for 2–5 min at 650 rpm.

9. Add 50 μl of RNase-free water to dried beads; shake for 3 min and then place on a magnetic stand for 1 min before aspirating 45 μl of eluted RNA, being careful not to disturb the beads. RNA concentration can be measured from a subset of wells to assess RNA concentration, consistency, and quality.

3.3 Synthesis of cDNA from Purified RNA

cDNA can be prepared using any reverse transcriptase. We use the Applied Biosystems (cat no. 4374967) using the following method-.

1. For each 96-well plate of RNA prepare a master mix containing:

   100 μl 10× RT buffer.

   40 μl 25× dNTPs.

   100 μl Random primers.

   50 μl Reverse Transcriptase.

   50 μl RNase inhibitor.

   160 μl RNase-free water.

2. Distribute 5 μl of the master mix and 5 μl of RNA to each well of the 96-well plate and centrifuge for 30 s at 2000 rpm to ensure that the components are well mixed.

3. Place the plate in a thermocycler and cycle through the following steps:

   25 °C for 10 min.

   37 °C for 120 min.

   85 °C for 5 min.

   Hold at 4 °C.

3.4 qPCR and Analysis Using the ΔΔCt Method for Relative Quantification

Quantitative measurement of RNA is best made using Real-time PCR, using single-color intercalating dyes like SYBR-green or a dual-color probe system like Taqman assays. RNA quantification using Taqman probes with multiplexing is described below.

3.4.1 qPCR

1. For each 96-well qPCR plate make a master mix containing the following components:

   53 μl 20× gene-specific TaqMan probe assay.

   53 μl 20× endogenous control TaqMan assay such as GAPDH.

   530 μl 2× Roche qPCR master mix, or other polymerase appropriate for Real Time PCR machine being used for amplification.

   330 μl purified water.

2. Distribute 8 μl of master mix to each well of a 384-well qPCR plate and centrifuge at 2000 rpm for 15 s.

3. Add 2 μl of cDNA to the master mix, seal plates with transparent film that is appropriate for your qPCR machine, and centrifuge as above.

4. Run qPCR (Fig. 2) using an appropriate method to detect fluorescent signals from gene-specific and endogenous control TaqMan probes. A typical set uses FAM and VIC combination using Dual Color Hydrolysis/UPL Probe program on a Roche LightCycler 480.

3.4.2 Analysis of qPCR Using Relative Quantification

Typically, the ΔΔCt method of relative quantification is utilized to quantify gene expression agnostic to qPCR platform.

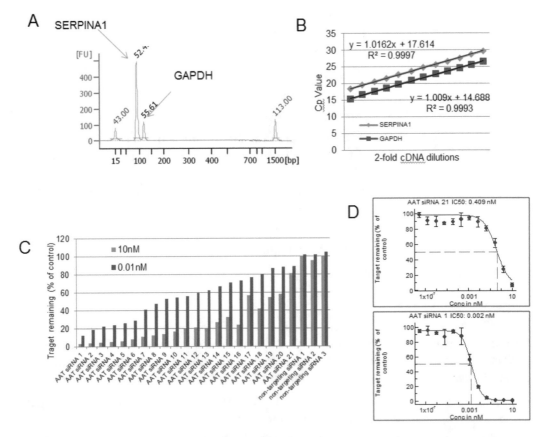

Fig. 2 TaqMan assay validation and in vitro screening data. (**a**) TaqMan assays validated by amplicon size confirmation showing individual peaks for SERPINA1 and GAPDH, using Agilent 2100 Bioanalyzer (**b**) and linearity of amplification of SERPINA1 (*blue line*) and GAPDH (*red line*) over a range of cDNA dilutions. (**c**) Activity for a series of AAT siRNAs screened at two doses (10 nM *blue*, 0.01 nM *red*) and (**d**) dose response screens from the most and least active siRNAs. *Red dotted line* defines IC50 value extrapolated from a four-parameter fit model. $n = 4$

1. Follow the method for export of threshold values from your qPCR machine (called Ct by Applied Biosystems or Cp by Roche) for the target gene and endogenous control. The Cp terminology will be used here.

2. Subtract the Cp value of the endogenous control from the Cp value of the target gene to determine the ΔCp. This step will normalize values within each well to control for differences in cDNA input.

3. Average ΔCp values from nontargeting negative controls or naive wells.

4. Subtract the negative control ΔCp from the target ΔCp to generate the ΔΔCp.

5. To determine fold change use the following formula: $2^{\wedge}(\Delta\Delta Cp)$. This value can be multiplied by 100 to determine the percent message remaining relative to the negative control.

3.5 In Vivo Analysis of siRNA for AAT Knockdown

The inherent potency of siRNA in reducing AAT mRNA levels is tested in cell lines. The next step is to test the ability of these molecules to silence genes in vivo (in animals). The most potent siRNA from in vitro screens are chosen for the next round of selection in animals. The efficacy of compounds can be measured at both the mRNA and the circulating protein level. mRNA measurements in rodents require the sacrifice of animals to collect the target organ, in this case, the liver. Protein measurements in blood can be made by immunological methods using antibodies (e.g., western blots or ELISA), or by measuring activity of the target protein.

The protocol below describes an example of siRNA screened in rats. These sequences were picked from the in vitro set described above. All animal studies are conducted in accordance with animal welfare regulations under IACUC-approved research protocols.

Experimental design: $n = 4$, female rats

1. Collect pre-dose bleeds from rats as per the IAUCAC protocols. Let the blood clot at room temperature for 30–45 min, spin and collect the serum. Aliquot appropriately and freeze.

2. Dilute siRNA aseptically in 1× PBS or normal saline to allow for appropriate injection volumes. While comparing compounds, it is best to inject molecules at the same concentration.

3. Inject anaesthetized animals at their back subcutaneously. Ensure that there is no leakage of material from the injection site.

4. Sacrifice animals on day 7, collect blood in serum collection tubes and a liver lobe/piece in appropriate jar/tube. Flash freeze the liver until ready to process.

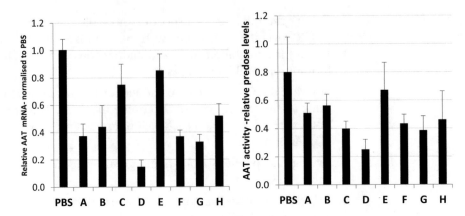

Fig. 3 Female rats ($n = 4$) were injected subcutaneously with PBS or siRNA (A_H) (at 2 mg/kg) on Day 0. Animals were sacrificed on Day 7. A: Relative AAT mRNA, normalized to GAPDH. B: AAT activity, relative to pre-dose levels. Error bars represent standard deviation for each group

5. Isolate liver RNA, reverse transcribe to make cDNA, and measure the amount of residual mRNA as described above (Fig. 3a) [17].

3.6 AAT Activity Assay

AAT activity in rat serum is measured using a Neutrophil Elastase Inhibitor Screening Kit (cat no. ab118971). This assay is based on the ability of AAT to inhibit the serine protease, neutrophil elastase, that can cleave a substrate to generate a fluorescence signal from substrate hydrolization that can be detected at EX/EM = 400/505 nm.

1. Mix 25 μl of diluted rat serum (10,000–20,000-fold dilution in assay buffer) with 50 μl of human elastase. A serial dilution of purified AAT at designated concentration is also mixed with elastase to generate a standard curve.

2. Incubate at 37 °C for 10 min protected from light.

3. Add 25 μl of substrate N-(Methoxysuccinyl)-Ala-Ala-Pro-Val-7-amino-4-trifluormethylcoumarin to the reaction. Quickly mix.

4. Measure RFU immediately (EX/EM = 400/505 nm) at 37 °C in a 30-s interval for 5 min using SpectraMax 5e.

5. Reaction rate is calculated as $\text{rate(RFU/s)} = \frac{\text{RFU}(300s) - \text{RFU}(0s)}{300s}$

Rate is then fit back to the standard curve to obtain AAT activity in serum. The relative activity in experimental groups is normalized to the PBS group (Fig. 3b).

The best molecule can be further characterized for dose response, duration of response, and impact of repeat doses.

There is little similarity between mouse and human SerpinA1 (AAT) at the nucleotide level, so a transgenic mouse model expressing human Z-AAT (Tg-PiZ) is used to screen siRNA against human siRNA. Phenotypic consequences of target silencing can also be evaluated in Tg-PiZ mice. A typical mouse screening study is described below:-

1. Collect pre-dose ocular bleeds from the Tg-PiZ mice in serum collector tubes.

2. Dilute siRNA aseptically with $1\times$ PBS or normal saline, e.g., to 0.1 mg/ml to allow for dosing at 1 mg/kg.

3. Inject animals subcutaneously at the back, at 10 μl/g bodyweight, as per IACUC protocols on Day 0.

4. Collect ocular bleeds at appropriate days to capture timecourse and duration of action of the siRNA, e.g., on days 3, 7, 11, 15, 21, 27.

5. Measure human AAT using an ELISA that specifically recognizes human AAT. For example, the Abcam ELISA (cat no. ab108799), with the Tg-PiZ mouse serum diluted at 1:50,000 dilution.

6. Each Tg-PiZ mouse has slightly different amounts of circulating hAAT, so the results are best represented after normalization to the pre-dose AAT levels. Figure 4 shows an example where five compounds were screened for activity in Tg-PiZ mice. Compound E shows maximum potency and duration in this test and would be selected for further evaluation.

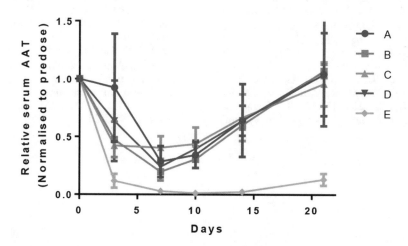

Fig. 4 Multiple siRNA were diluted in $1\times$ PBS and dosed at 10 mg/kg. Ocular bleeds were collected before dosing, and on days 3, 7, 10, 14, and 21. Serum samples were analyzed for AAT levels using ELISA. Each plotted data point represents the mean of normalized AAT from three animals; error bars represent the standard deviation

References

1. Miravitlles M (2012) Alpha-1-antitrypsin and other proteinase inhibitors. Curr Opin Pharmacol 12(3):309–314. doi:10.1016/j.coph.2012.02.004. S1471-4892(12)00027-6 [pii]

2. Stockley RA, Turner AM (2014) Alpha-1-antitrypsin deficiency: clinical variability, assessment, and treatment. Trends Mol Med 20(2):105–115. doi:10.1016/j.molmed.2013.11.006

3. Stockley RA (2014) Alpha1-antitrypsin review. Clin Chest Med 35(1):39–50. doi:10.1016/j.ccm.2013.10.001

4. Salahuddin P (2010) Genetic variants of alpha1-antitrypsin. Curr Protein Pept Sci 11(2):101–117. doi:CPPS-23 [pii]

5. Stoller JK, Aboussouan LS (2005) Alpha1-antitrypsin deficiency. Lancet 365(9478):2225–2236. doi:10.1016/s0140-6736(05)66781-5

6. Gooptu B, Dickens JA, Lomas DA (2014) The molecular and cellular pathology of alpha(1)-antitrypsin deficiency. Trends Mol Med 20(2):116–127. doi:10.1016/j.molmed.2013.10.007

7. Wu Y, Whitman I, Molmenti E, Moore K, Hippenmeyer P, Perlmutter DH (1994) A lag in intracellular degradation of mutant alpha 1-antitrypsin correlates with the liver disease phenotype in homozygous PiZZ alpha 1-antitrypsin deficiency. Proc Natl Acad Sci U S A 91(19):9014–9018

8. Elbashir SM, Lendeckel W, Tuschl T (2001) RNA interference is mediated by 21- and 22-nucleotide RNAs. Genes Dev 15(2):188–200

9. Soutschek J, Akinc A, Bramlage B, Charisse K, Constien R, Donoghue M, Elbashir S, Geick A, Hadwiger P, Harborth J, John M, Kesavan V, Lavine G, Pandey RK, Racie T, Rajeev KG, Rohl I, Toudjarska I, Wang G, Wuschko S, Bumcrot D, Koteliansky V, Limmer S, Manoharan M, Vornlocher HP (2004) Therapeutic silencing of an endogenous gene by systemic administration of modified siRNAs. Nature 432(7014):173–178

10. Ashwell G, Morell AG (1974) The role of surface carbohydrates in the hepatic recognition and transport of circulating glycoproteins. Adv Enzymol Relat Areas Mol Biol 41(0):99–128

11. Baenziger JU, Fiete D (1980) Galactose and N-acetylgalactosamine-specific endocytosis of glycopeptides by isolated rat hepatocytes. Cell 22(2 Pt 2):611–620

12. Schwartz AL, Rup D, Lodish HF (1980) Difficulties in the quantification of asialoglycoprotein receptors on the rat hepatocyte. J Biol Chem 255(19):9033–9036

13. Geffen I, Spiess M (1992) Asialoglycoprotein receptor. Int Rev Cytol 137B:181–219

14. Wu J, Nantz MH, Zern MA (2002) Targeting hepatocytes for drug and gene delivery: emerging novel approaches and applications. Front Biosci 7:d717–d725

15. Nair JK, Willoughby JL, Chan A, Charisse K, Alam MR, Wang Q, Hoekstra M, Kandasamy P, Kel'in AV, Milstein S, Taneja N, O'Shea J, Shaikh S, Zhang L, van der Sluis RJ, Jung ME, Akinc A, Hutabarat R, Kuchimanchi S, Fitzgerald K, Zimmermann T, van Berkel TJ, Maier MA, Rajeev KG, Manoharan M (2014) Multivalent N-acetylgalactosamine-conjugated siRNA localizes in hepatocytes and elicits robust RNAi-mediated gene silencing. J Am Chem Soc 136(49):16958–16961. doi:10.1021/ja505986a

16. Milstein S, Nguyen M, Meyers R, de Fougerolles A (2013) Measuring RNAi knockdown using qPCR. Methods Enzymol 533:57–77. doi:10.1016/B978–0–12-420067-8.00006-4. B978-0-12-420067-8.00006-4 [pii]

17. Sehgal A, Chen Q, Gibbings D, Sah DW, Bumcrot D (2014) Tissue-specific gene silencing monitored in circulating RNA. RNA 20(2):143–149. doi:10.1261/rna.042507.113. rna.042507.113 [pii]

Chapter 12

Knockdown of Z Mutant Alpha-1 Antitrypsin In Vivo Using Modified DNA Antisense Oligonucleotides

Mariam Aghajan, Shuling Guo, and Brett P. Monia

Abstract

Alpha-1 antitrypsin (AAT) is a serum protease inhibitor, mainly expressed in and secreted from hepatocytes, important for regulating neutrophil elastase activity among other proteases. Various mutations in AAT cause alpha-1 antitrypsin deficiency (AATD), a rare hereditary disorder that results in liver disease due to accumulation of AAT aggregates and lung disease from excessive neutrophil elastase activity. PiZ transgenic mice contain the human *AAT* genomic region harboring the most common AATD mutation, the Glu342Lys (Z) point mutation. These mice effectively recapitulate the liver disease exhibited in AATD patients, including AAT protein aggregates, hepatocyte death, and eventual liver fibrosis. Previously, we demonstrated that modified antisense oligonucleotides (ASOs) can dramatically reduce Z-AAT RNA and protein levels in PiZ mice enabling inhibition, prevention, and reversal of the associated liver disease. Here, we describe in detail usage of AAT-ASOs to knock down Z-AAT in PiZ mice with a focus on preparation and in vivo delivery of ASOs, as well as detailed workflows pertaining to the analysis of Z-AAT mRNA, plasma protein, and soluble/insoluble liver protein levels following ASO administration.

Key words Alpha-1 antitrypsin, Z-AAT, PiZ mice, Chemically modified antisense oligonucleotides (ASO), In vivo delivery, Knockdown, AAT mRNA, AAT protein, AAT Western blot

1 Introduction

Alpha-1 antitrypsin (AAT), part of the serpin super family, is the most abundant serine protease inhibitor in circulation. One of its main targets is neutrophil elastase that when excessively active is responsible for the destruction of lung connective tissue [1]. Mutations in *AAT* cause alpha-1 antitrypsin deficiency (AATD), a rare genetic disease that has two distinct manifestations: (1) adult-onset emphysema due to decreased AAT activity and uncontrolled neutrophil elastase activity and (2) liver disease due to aggregation and retention of mutant AAT in hepatocytes [2–9]. Although over 100 mutations in AAT have been identified, more than 90% of AATD patients are homozygous for the Z (Glu342Lys) point mutation (PiZZ patients), which causes the most severe liver and lung

Florie Borel and Christian Mueller (eds.), *Alpha-1 Antitrypsin Deficiency: Methods and Protocols*, Methods in Molecular Biology, vol. 1639, DOI 10.1007/978-1-4939-7163-3_12, © Springer Science+Business Media LLC 2017

disease. This mutation results in a "loop-sheet" conformation that causes insoluble Z-AAT protein aggregates to accumulate in the rough ER in hepatocytes, preventing protein secretion and most likely triggering liver injury [10, 11]. The onset of disease is varied in PiZZ patients with some exhibiting symptoms of neonatal jaundice and cholestasis in infancy/early childhood, while others develop slow progressing fibrosis typically diagnosed in their 5th decade [12, 13]. Regardless, subsets of these patients go on to develop advanced fibrosis, cirrhosis, and/or hepatocellular carcinoma. Although AAT protein replacement therapy is available as treatment for AATD lung disease, liver transplantation is the only treatment option for AATD liver disease.

In an effort to characterize the liver injury caused by Z-AAT in PiZZ patients, PiZ transgenic mice have been generated using the human *AAT* genomic region harboring the Z point mutation [14]. This transgene is driven from its endogenous promotor, and as such, Z-AAT expression in these mice closely resembles patterns observed in human tissues with expression detected primarily in hepatocytes [15]. Since PiZ mice retain expression of endogenous wild-type mouse AAT, these mice do not exhibit the lung disease component of AATD [14]. However, PiZ mice effectively recapitulate the liver disease presented in PiZZ patients, exhibiting AAT insoluble protein aggregates in the ER of hepatocytes and liver injury that increases as Z-AAT aggregates accumulate. Moreover, in PiZZ (homozygous) mice, liver fibrosis is established by 4 months of age demonstrating that the PiZ mouse model is an appropriate in vivo model for translational therapeutic studies [16–18].

Chemically modified antisense oligonucleotides (ASOs) can be used to deplete specific RNAs in vivo, providing a valuable tool for studying the role of any RNA (or its resultant protein) in a physiological or pathophysiological context. We have successfully applied this technology toward targeting a number of different transcripts in a wide range of rodent disease models, primates, and humans. ASOs are single-stranded and typically comprised of 14–20 DNA nucleotides that consist of unmodified nucleotides in the middle flanked by chemically modified nucleotides in the wings (the so-called Gapmer ASO design). Currently, we use two different chemistries to modify the sugar moiety of nucleotides in the wings, both of which result in increased affinity, nuclease resistance, and tolerability: 2′-O-methoxyethyl (MOE) and constrained ethyl bicyclic nucleic acid (cET) [19]. Moreover, the ASO backbone is altered with phosphorothioate linkages that improve the pharmacokinetic properties due to enhanced nuclease resistance and increased ASO binding to plasma proteins. ASOs bind their RNA targets by Watson-Crick hybridization forming a DNA-RNA heteroduplex that is recognized by the ubiquitously expressed RNase-H enzyme causing cleavage of the bound RNA [20].

Previously, we reported that administering hAAT ASOs to PiZ mice dramatically reduces Z-AAT liver mRNA and protein levels, as well as circulating Z-AAT protein levels [21, 22]. These reductions were sufficient to prevent, inhibit, and reverse Z-AAT aggregate accumulation and liver injury in this model. Moreover, we demonstrated inhibition of liver fibrosis with ASO treatment. Here, we describe how hAAT ASOs can be used to achieve therapeutic efficacy in the PiZ mouse model. As in silico design and synthesis of ASOs have been described previously [23], we begin our workflow with preparation of ASOs and in vivo ASO delivery, followed by steps taken to analyze Z-AAT liver mRNA, plasma protein, and soluble/insoluble liver protein knockdown following ASO treatment.

2 Materials and Equipment

2.1 Preparation of ASOs

1. D-PBS, no calcium, no magnesium (Life Technologies).
2. 0.22 μm Spin-X centrifuge tube filters, cellulose acetate membrane (Corning).
3. Tris-EDTA buffer (10 mM Tris, 1 mM EDTA, pH 8.0 with HCl).
4. Spectrophotometer.

2.2 In Vivo Administration of ASOs

1. EDTA-treated blood collection tubes (BD).
2. Sterile liver biopsy punch (6 mm, Miltex).
3. Guanidine Isothiocyanate (GITC) (Life Technologies).
4. β-mercaptoethanol (Sigma).

2.3 Measuring Efficiency of Z-AAT RNA Knockdown

1. β-mercaptoethanol (Sigma).
2. RNeasy 96 kit (Qiagen).
3. StepOne Plus Real-Time PCR machine (Applied Biosystems).
4. EXPRESS One-Step SuperScript qRT-PCR Universal kit (Life Technologies).
5. Quant-iT RiboGreen RNA Assay kit (Life Technologies).

2.4 Measuring Efficiency of Circulating Z-AAT Protein Knockdown

1. D-PBS, no calcium, no magnesium (Life Technologies).
2. AU480 Clinical Chemistry Analyzer (Beckman Coulter).
3. Alpha-1 Antitrypsin reagent set (DiaSorin).

2.5 Measuring Effects of ASO Administration on Soluble and Insoluble Z-AAT Protein Levels in Liver

1. Glass dounce homogenizer.

2. cOmplete, mini, EDTA-free protease inhibitor cocktail tablets (Roche).

3. Buffer A: 50 mM Tris–HCl pH 8.0, 150 mM NaCl, 5 mM KCl, 5 mM $MgCl_2$, 0.5% Triton X-100, and 1 tablet cOmplete mini EDTA-free protease inhibitors for every 10 mL Buffer A.

4. 27 ½-guage needles.

5. 1 mL syringes.

6. BCA Protein Assay kit (Pierce).

7. 4× LDS (Life Technologies).

8. β-mercaptoethanol (Sigma).

9. Buffer B: D-PBS containing 1% Triton X-100, 0.05% Na-deoxycholate, 10 mM EDTA, and 1 tablet cOmplete mini EDTA-free protease inhibitors for every 10 mL Buffer B.

10. Sonicator.

11. NuPAGE Novex 4–12% Bis-Tris protein gels, 1.0 mm (Life Technologies).

12. 20× MOPS buffer (Life Technologies).

13. Polyclonal human Alpha-1 Antitrypsin antibody (DiaSorin).

14. HRP-conjugated donkey anti-goat secondary antibody (Jackson ImmunoResearch).

15. ECL Prime Western blotting detection kit (GE Amersham).

3 Methods

3.1 Preparation of ASOs

1. Prepare ASO stock solution by dissolving ASO powder with D-PBS to a concentration of ~100 mg/mL (*see* **Note 1**).

2. Filter ASO stock solution using a 0.22 μm Spin-X filter column and centrifuging at > 10,000 × *g* for 2 min or until all ASO has filtered through (*see* **Note 2**).

3. Dilute ASO stock solution with TE buffer and measure OD_{260} using a spectrophotometer to determine ASO concentration (*see* **Notes 3** and **4**).

4. Prepare ASO working solution by diluting ASO stock solution with D-PBS to a concentration of 2.5 mg/mL.

5. Recalculate ASO concentration by diluting ASO working solution with TE buffer and measuring OD_{260} (*see* **Note 5**).

6. Store ASO solutions at 4 °C for short-term storage (i.e., ASO working solution for duration of study) and at −20 °C for long-term storage (i.e., ASO stock solution).

3.2 In Vivo Administration of ASOs

ASOs can be delivered systemically to animals via subcutaneous (SC), intravenous (IV), or intraperitoneal (IP) injection. These routes of administration provide broad distribution of ASOs in vivo with most efficient delivery to liver and kidney, followed by muscle, heart, colon, and fat [24, 25]. Moreover, ASOs can be delivered locally to the eye via intravitreal (IVT) injection [26, 27], CNS via intrathecal (IT) or intracerebroventricular (ICV) injection [28, 29], and lung via aerosol delivery [30]. To target hAAT, we used MOE ASOs and administered them to PiZ mice via subcutaneous injection; however, similar results can be obtained with other means of systemic ASO administration.

The age of PiZ mice and the duration of ASO treatment required depend on the objective of the study. To inhibit progression of AATD liver disease in PiZ mice, we administered ASOs to 6-week-old male PiZ mice for 8 weeks [21]. To demonstrate that ASO treatment can prevent liver disease in these mice, we treated 2-week-old male PiZ mice, which have small Z-AAT liver aggregates at this age, with ASO for a duration of 8 weeks. AATD liver disease reaches steady state in PiZ mice by 4 months of age at which point liver globule accumulation is near maximum. Therefore, we administered 16-week-old male PiZ mice with ASO for 20 weeks to demonstrate the ability of ASO treatment to reverse insoluble aggregate accumulation and liver injury in this model. Importantly, all studies included the proper controls such as a vehicle-treated (D-PBS only) group and a control ASO group. Control ASO is a scrambled ASO comprised of the same chemistry as that of the test ASO accounting for ASO-specific effects unrelated to pharmacological effects of the ASO. Additionally, it is critical to sacrifice a group of age-matched mice at the start of the study to have "baseline" samples with which to compare at the end of the study in order to properly assess ASO effects on disease progression.

1. Weigh mice prior to first dose and weekly thereafter (*see* **Note 6**).

2. Administer MOE hAAT ASO and control ASO (2.5 mg/mL) via subcutaneous injection twice weekly at 10 μL per gram body weight which equates to 25 mg/kg per dose (*see* **Notes 7** and **8**).

3. To assess effects of ASO treatment on circulating levels of Z-AAT during course of study, collect 200–250 μL whole blood from each mouse by tail nick into EDTA-treated blood collection tubes and gently mix tube by inverting before placing on ice. Spin down samples at $>10,000 \times g$ for 8 min and remove supernatant plasma to new tubes. Store plasma samples at $-80\ °C$ until end of study when samples can be assayed simultaneously and avoid repeated freeze-thaws to minimize Z-AAT protein degradation.

4. Sacrifice mice 48 h after the last dose and collect two liver punches from each animal using a liver biopsy punch. Snap-freeze one liver punch in liquid nitrogen using cryogenic vials to be used for protein analysis; homogenize a second liver punch for 20 s in 14 mL falcon tubes containing 2 mL GITC supplemented with 8% β-mercaptoethanol to be used for RNA analysis. Transfer homogenized liver lysates to a 96-well plate and store both snap-frozen liver punches and homogenized tissue lysates at −80 °C.

3.3 Measuring Efficiency of Z-AAT RNA Knockdown

1. Transfer 30 μL GITC/β-mercaptoethanol liver lysate to a 96-well plate and add 70 μL Qiagen RLT buffer containing β-mercaptoethanol to each sample to bring the total volume to 100 μL. Transfer two sets of lysates from PBS-treated samples such that RNA from one set can be used for generating standard curve samples.

2. Seal plate and mix well.

3. Spin down the plate and extract RNA following the Qiagen RNeasy 96 kit protocol.

4. Elute RNA with 150 μL RNase-free water.

5. Pool samples extracted for standard curve generation into separate Eppendorf tube and perform two-fold serial dilutions with RNase-free water for five dilutions.

6. Perform 1:1 dilution of remaining purified RNA by adding 120 μL RNase-free water to a new 96-well plate and transferring 120 μL purified RNA to the new plate.

7. Seal plate and mix well.

8. To measure levels of Z-AAT RNA knockdown, conduct qRT-PCR (TaqMan) using 5 μL purified RNA in a 20 μL reaction volume using the One-Step SuperScript qRT-PCR kit. Mouse AAT levels may also be assessed to ensure specificity of ASOs (*see* **Note 9**) (Fig. 1).

9. To normalize RNA expression levels, use the Quant-iT RiboGreen RNA assay kit to measure relative RNA levels (*see* **Note 10**).

3.4 Measuring Efficiency of Circulating Z-AAT Protein Knockdown

A clinical chemistry analyzer is a reliable and efficient instrument capable of measuring various plasma, serum, and urine analytes using an immunoturbidity method. The human AAT (hAAT) assay available from Beckman Coulter is an appropriate assay to measure plasma AAT from various species and has been used effectively to measure monkey AAT levels [21]. Due to the strong cross-reactivity of this assay with mouse AAT, however, a custom assay with DiaSorin Alpha-1 Antitrypsin reagent set was established and validated on the clinical analyzer where cross-reactivity with mouse AAT is minimized while maintaining sensitivity to both mutant (Z) and wild-type human AAT in the PiZ mouse samples.

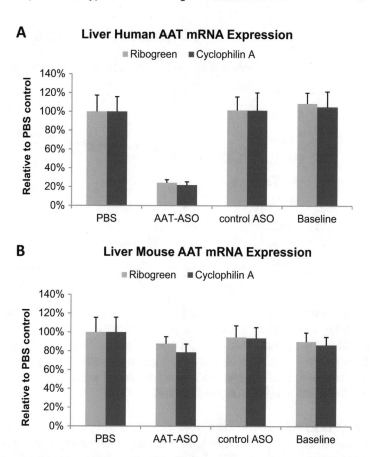

Fig. 1 Human *AAT* mRNA levels are reduced in PiZ mouse livers after AAT-ASO treatment. 6-week-old male PiZ mice were treated with PBS, 25 mg/kg AAT-ASO or control ASO twice a week for 8 weeks via subcutaneous injection. A group of age-matched mice were sacrificed prior to the start of ASO treatment for baseline analysis. Liver biopsies from each mouse were collected 2 days following the last dose and (**a**) human *AAT* and (**b**) mouse *AAT* mRNA levels were assessed via qRT-PCR. Normalization of mRNA was performed using either Ribogreen values or relative Cyclophilin A mRNA expression levels, both producing similar results. Results represent means $\pm$ standard deviations ($n = 4$)

1. Thaw all plasma samples collected during course of study at the same time to limit variations between samples.

2. Dilute plasma samples $11\times$ with D-PBS and vortex to mix well (i.e., 10 μL plasma +100 μL D-PBS).

3. Measure plasma AAT levels on an AU480 Clinical Chemistry Analyzer without entering the dilution factor using DiaSorin AAT reagent.

4. Since the DiaSorin assay has weak cross-reactivity with mouse AAT, hAAT levels can be determined by subtracting the values obtained for AAT by 9.25, which accounts for endogenous mouse AAT levels (*see* **Note 11**) (Fig. 2).

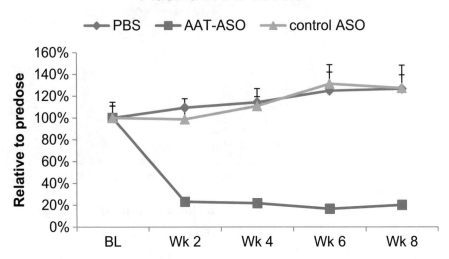

Fig. 2 Human AAT plasma protein levels are reduced in PiZ mice after AAT-ASO treatment. 6-week-old male PiZ mice were treated with PBS, 25 mg/kg AAT-ASO or control ASO twice a week for 8 weeks via subcutaneous injection. Mice were tail bled for plasma collection prior to initiation of ASO treatment to determine baseline hAAT levels and every 2 weeks thereafter through end of study. Human AAT levels were measured using a clinical chemistry analyzer and data is presented with respect to individual baseline levels. Results represent means ± standard deviations ($n = 4$)

3.5 Measuring Effects of ASO Administration on Soluble and Insoluble Z-AAT Protein Levels in Liver

Insoluble Z-AAT aggregates accumulate in livers of PiZ mice as they age, reaching near steady state by 4 months of age [21]. In PiZZ (homozygous) mice, however, liver injury is more severe such that hepatocytes containing aggregates undergo apoptosis and are cleared from the liver. As a result, ASO administration in PiZZ mice fails to show an effect on liver globule formation despite proving to be efficacious against liver fibrosis in this model. Therefore, PiZ mice are more appropriate for assessing the effects of ASO treatment on Z-AAT insoluble aggregates in liver. The protocol outlined herein is modified from a previously described method [31].

1. Cut frozen liver punch using frozen razor blade and forceps to isolate ¼ liver punch for protein extraction.

2. Transfer ¼ liver punch to ice-cold dounce homogenizer on ice and add 2 mL ice-cold Buffer A to the sample.

3. Homogenize tissue on ice using Pestle A (loose) for 30–40 repetitions, then transfer 1 mL of lysate to new chilled eppendorf tubes, and vortex (*see* **Note 12**).

4. Pass 1 mL lysate through a 27 ½-guage needle ten times using chilled syringes.

5. Measure total protein concentration of lysate using BCA Protein Assay kit (*see* **Note 13**).

6. Aliquot 150 µg protein lysate into new chilled Eppendorf tubes for fractionation of insoluble and soluble proteins and bring total volume of lysate to 300 µL with Buffer A (continued at **step 8**).

7. For total protein samples, aliquot 50 µg protein lysate into a separate set of chilled Eppendorf tubes and bring total volume to 130 µL with Buffer A. Add 50 µL 4× LDS and 20 µL β-mercaptoethanol to lysates. Boil samples at 70 °C for 10 min and store at −20 °C until ready to load gel.

8. For fractionation of soluble and insoluble proteins, centrifuge 150 µg protein lysates from **step 6** at 10,000 × *g* for 30 min at 4 °C.

9. Remove 100 µL supernatant to new chilled Eppendorf tubes (soluble protein fraction) and add 30 µL Buffer A, 50 µL 4× LDS, and 20 µL β-mercaptoethanol for a total volume of 200 µL. Boil samples at 70 °C for 10 min and store at −20 °C until ready to load gel.

10. Discard remaining supernatant from pelleted samples from **step 8** and dissolve pellet with 300 µL ice-cold Buffer B (insoluble protein fraction) by vortex (*see* **Note 14**).

11. Sonicate samples on ice for 10 s using 10% amplitude and then vortex.

12. Add 135 µL 4× LDS, 54 µL β-mercaptoethanol, and 51 µL Buffer B for a total volume of 540 µL. Boil samples at 70 °C for 10 min and store at −20 °C until ready to load gel.

13. For Western blot analysis run samples on 4–12% Bis-Tris NuPAGE gel using MOPS buffer.

14. Load 15 µL total and soluble protein samples and 13.5 µL insoluble protein sample per lane to ensure equal amounts of protein loaded per lane (3.75 µg per lane).

15. Run gel at 150 V for 1 h, 40 min and transfer gel to membrane.

16. Block membrane with 5% milk in TBST for 1 h at RT.

17. Wash membrane 3× with TBST for 5 min each at RT and probe for hAAT protein overnight at 4 °C using DiaSorin AAT antibody diluted at 1:20,000 in 5% milk/TBST.

18. Wash membrane 3× with TBST for 5 min each at RT and incubate with anti-goat secondary antibody diluted 1:40,000 in 5% milk/TBST for 1 h at RT.

19. Wash membrane 3× with TBST for 5 min each at RT and develop blot using ECL Prime Western blotting detection kit.

20. Blot for housekeeping protein, such as GAPDH, to ensure even protein loading (Fig. 3).

4 Notes

1. Typically, ASOs are left to dissolve overnight at RT on benchtop followed by vortexing, but if pressed for time, a couple of hours with intermittent vortexing should suffice.

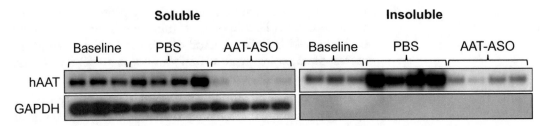

Fig. 3 AAT-ASO treatment reduces both soluble and insoluble liver hAAT proteins in PiZ mice. 8-week-old male PiZ mice were treated with PBS or 25 mg/kg AAT-ASO twice a week for 20 weeks via subcutaneous injection. A group of age-matched mice were sacrificed prior to the start of ASO treatment for baseline analysis. Liver biopsies from each mouse were collected 2 days following the last dose and snap-frozen for protein analysis. Liver tissue was lysed and soluble and insoluble protein fractions were isolated

2. Syringe filtration using 0.22 μm filters can also be used to filter ASO solutions.

3. Typically, a dilution factor of 3000 is used to obtain OD values in the linear range.

4. ASO concentration is calculated using the following equation: Concentration (mg/mL) = $(OD_{260} \times$ Dilution Factor $\times$ MW)/(Mass extinction coefficient $\times$ 1000).

5. At this concentration, a dilution factor of 500 is sufficient to obtain OD values in the linear range.

6. Groups of four mice per treatment are usually sufficient to achieve statistical significance.

7. Typically, ASOs are administered twice weekly at 25 mg/kg/dose. Similar results, however, can be obtained with once a week ASO administration at 50 mg/kg/dose. Additionally, these doses are appropriate for MOE ASOs; if cET ASOs are to be used, then dosing will need to adjusted accordingly as cET ASOs are more potent than MOE ASOs.

8. It is critical to measure blood chemistries at the conclusion of an ASO-administered study to ensure that ASOs are well-tolerated. Important blood parameters to assess include ALT (alanine transaminase), AST (aspartate transaminase), BUN (blood urea nitrogen), and total bilirubin levels.

9. Primer and probe concentrations of 10 μM and 2.5 μM, respectively, are typically used for qRT-PCR reactions. The sequences of primers and probes used to measure Z-AAT and mouse AAT levels are as follows: Z-AAT, forward, 5′-GGAGATGCTGCC-CAGAAGAC-3′, reverse, 5′-GCTGGCGGTATAGGCTGAA G-3′, probe, 5′-Fam-ATCAGGATCACCCAACCTTCAACA AGATCA-Tamra-3′; mouse AAT, forward, 5′-TTCTGGCA GGCCTGTGTTG-3′, reverse, 5′-ATCCTTCTGGGAGGTG TCTGTCT-3′, probe, 5′-Fam-CCCCAGCTTTCTGGCT-GAGGATGTTC-Tamra-3′.

10. RNA levels can also be normalized to relative expression levels of a housekeeping gene such as Cyclophilin A. The sequences of primers and probe used to measure mouse Cyclophilin A levels are as follows: forward, 5′-TCGCCGCTTGCTGCA-3′, reverse, 5′-ATCGGCCGTGATGTCGA-3′, probe, 5′-Fam-CCATGGTCAACCCCACCGTGTTC-Tamra-3′.

11. The correction value of 9.25 was calculated by measuring endogenous mAAT levels of normal mice using the DiaSorin AAT assay.

12. It is critical to homogenize sample sufficiently to remove all clumps otherwise, lysate will clog syringe in subsequent steps. Similarly, using too much tissue for homogenization can also result in insufficient homogenization and clogging of syringe.

13. Diluting lysates 4× with water is sufficient to measure protein concentration in the linear range of the assay.

14. Do not disturb pellet when removing the supernatant.

References

1. Janciauskiene SM, Bals R, Koczulla R, Vogelmeier C, Kohnlein T, Welte T (2011) The discovery of alpha1-antitrypsin and its role in health and disease. Respir Med 105 (8):1129–1139. doi:10.1016/j.rmed.2011.02.002

2. Bals R (2010) Alpha-1-antitrypsin deficiency. Best Pract Res Clin Gastroenterol 24 (5):629–633. doi:10.1016/j.bpg.2010.08.006

3. Ekeowa UI, Gooptu B, Belorgey D, Hagglof P, Karlsson-Li S, Miranda E, Perez J, MacLeod I, Kroger H, Marciniak SJ, Crowther DC, Lomas DA (2009) Alpha1-antitrypsin deficiency, chronic obstructive pulmonary disease and the serpinopathies. Clin Sci (Lond) 116 (12):837–850. doi:10.1042/CS20080484

4. Fregonese L, Stolk J (2008) Hereditary alpha-1-antitrypsin deficiency and its clinical consequences. Orphanet J Rare Dis 3:16. doi:10.1186/1750-1172-3-16

5. Gooptu B, Ekeowa UI, Lomas DA (2009) Mechanisms of emphysema in alpha1-antitrypsin deficiency: molecular and cellular insights. Eur Respir J 34(2):475–488. doi:10.1183/09031936.00096508

6. Greene CM, Miller SD, Carroll T, McLean C, O'Mahony M, Lawless MW, O'Neill SJ, Taggart CC, McElvaney NG (2008) Alpha-1 antitrypsin deficiency: a conformational disease associated with lung and liver manifestations. J Inherit Metab Dis 31(1):21–34. doi:10.1007/s10545-007-0748-y

7. Hogarth DK, Rachelefsky G (2008) Screening and familial testing of patients for alpha 1-antitrypsin deficiency. Chest 133(4):981–988. doi:10.1378/chest.07-1001

8. Silverman EK, Sandhaus RA (2009) Clinical practice. Alpha1-antitrypsin deficiency. N Engl J Med 360(26):2749–2757. doi:10.1056/NEJMcp0900449

9. Teckman JH, Lindblad D (2006) Alpha-1-antitrypsin deficiency: diagnosis, pathophysiology, and management. Curr Gastroenterol Rep 8(1):14–20

10. Ekeowa UI, Freeke J, Miranda E, Gooptu B, Bush MF, Perez J, Teckman J, Robinson CV, Lomas DA (2010) Defining the mechanism of polymerization in the serpinopathies. Proc Natl Acad Sci U S A 107(40):17146–17151. doi:10.1073/pnas.1004785107

11. Lomas DA, Evans DL, Finch JT, Carrell RW (1992) The mechanism of Z alpha 1-antitrypsin accumulation in the liver. Nature 357(6379):605–607. doi:10.1038/357605a0

12. Fairbanks KD, Tavill AS (2008) Liver disease in alpha 1-antitrypsin deficiency: a review. Am J Gastroenterol 103(8):2136–2141.; quiz 2142. doi:10.1111/j.1572-0241.2008.01955.x

13. Nelson DR, Teckman J, Di Bisceglie AM, Brenner DA (2012) Diagnosis and management of patients with alpha1-antitrypsin (A1AT) deficiency. Clin Gastroenterol Hepatol 10(6):575–580. doi:10.1016/j.cgh.2011.12.028

14. Carlson JA, Rogers BB, Sifers RN, Finegold MJ, Clift SM, DeMayo FJ, Bullock DW, Woo SL (1989) Accumulation of PiZ alpha 1-antitrypsin causes liver damage in transgenic mice. J Clin Invest 83(4):1183–1190. doi:10.1172/JCI113999

15. Sifers RN, Carlson JA, Clift SM, DeMayo FJ, Bullock DW, Woo SL (1987) Tissue specific expression of the human alpha-1-antitrypsin gene in transgenic mice. Nucleic Acids Res 15 (4):1459–1475

16. Hidvegi T, Ewing M, Hale P, Dippold C, Beckett C, Kemp C, Maurice N, Mukherjee A, Goldbach C, Watkins S, Michalopoulos G, Perlmutter DH (2010) An autophagy-enhancing drug promotes degradation of mutant alpha1-antitrypsin Z and reduces hepatic fibrosis. Science 329(5988):229–232. doi:10.1126/science.1190354

17. Mencin A, Seki E, Osawa Y, Kodama Y, De Minicis S, Knowles M, Brenner DA (2007) Alpha-1 antitrypsin Z protein (PiZ) increases hepatic fibrosis in a murine model of cholestasis. Hepatology 46(5):1443–1452. doi:10.1002/hep.21832

18. Perlmutter DH, Silverman GA (2011) Hepatic fibrosis and carcinogenesis in alpha1-antitrypsin deficiency: a prototype for chronic tissue damage in gain-of-function disorders. Cold Spring Harb Perspect Biol 3(3):a005801. doi:10.1101/cshperspect.a005801

19. Bennett CF, Swayze EE (2010) RNA targeting therapeutics: molecular mechanisms of antisense oligonucleotides as a therapeutic platform. Annu Rev Pharmacol Toxicol 50:259–293. doi:10.1146/annurev.pharmtox.010909.105654

20. Crooke ST (1999) Molecular mechanisms of action of antisense drugs. Biochim Biophys Acta 1489(1):31–44

21. Guo S, Booten SL, Aghajan M, Hung G, Zhao C, Blomenkamp K, Gattis D, Watt A, Freier SM, Teckman JH, McCaleb ML, Monia BP (2014) Antisense oligonucleotide treatment ameliorates alpha-1 antitrypsin-related liver disease in mice. J Clin Invest 124 (1):251–261. doi:10.1172/JCI67968

22. Guo S, Booten SL, Watt A, Alvarado L, Freier SM, Teckman JH, McCaleb ML, Monia BP (2014) Using antisense technology to develop a novel therapy for alpha-1 antitrypsin deficient (AATD) liver disease and to model AATD lung disease. Rare Dis 2:e28511. doi:10.4161/rdis.28511

23. Zong X, Huang L, Tripathi V, Peralta R, Freier SM, Guo S, Prasanth KV (2015) Knockdown of nuclear-retained long noncoding RNAs using modified DNA antisense oligonucleotides. Methods Mol Biol 1262:321–331. doi:10.1007/978-1-4939-2253-6_20

24. Graham MJ, Lemonidis KM, Whipple CP, Subramaniam A, Monia BP, Crooke ST, Crooke RM (2007) Antisense inhibition of proprotein convertase subtilisin/kexin type 9 reduces serum LDL in hyperlipidemic mice. J Lipid Res 48(4):763–767. doi:10.1194/jlr.C600025-JLR200

25. Hung G, Xiao X, Peralta R, Bhattacharjee G, Murray S, Norris D, Guo S, Monia BP (2013) Characterization of target mRNA reduction through in situ RNA hybridization in multiple organ systems following systemic antisense treatment in animals. Nucleic Acid Ther 23 (6):369–378. doi:10.1089/nat.2013.0443

26. Geary RS, Henry SP, Grillone LR (2002) Fomivirsen: clinical pharmacology and potential drug interactions. Clin Pharmacokinet 41 (4):255–260. doi:10.2165/00003088-200241040-00002

27. Grillone LR, Lanz R (2001) Fomivirsen. Drugs Today (Barc) 37(4):245–255

28. Butler M, Hayes CS, Chappell A, Murray SF, Yaksh TL, Hua XY (2005) Spinal distribution and metabolism of 2′-O-(2-methoxyethyl)-modified oligonucleotides after intrathecal administration in rats. Neuroscience 131 (3):705–715. doi:10.1016/j.neuroscience.2004.11.038

29. Passini MA, Bu J, Richards AM, Kinnecom C, Sardi SP, Stanek LM, Hua Y, Rigo F, Matson J, Hung G, Kaye EM, Shihabuddin LS, Krainer AR, Bennett CF, Cheng SH (2011) Antisense oligonucleotides delivered to the mouse CNS ameliorate symptoms of severe spinal muscular atrophy. Sci Transl Med 3(72):72ra18. doi:10.1126/scitranslmed.3001777

30. Templin MV, Levin AA, Graham MJ, Aberg PM, Axelsson BI, Butler M, Geary RS, Bennett CF (2000) Pharmacokinetic and toxicity profile of a phosphorothioate oligonucleotide following inhalation delivery to lung in mice. Antisense Nucleic Acid Drug Dev 10 (5):359–368

31. An JK, Blomenkamp K, Lindblad D, Teckman JH (2005) Quantitative isolation of alpha1AT mutant Z protein polymers from human and mouse livers and the effect of heat. Hepatology 41(1):160–167. doi:10.1002/hep.20508

Chapter 13

Immunohistochemistry Staining for Human Alpha-1 Antitrypsin

Dongtao A. Fu and Martha Campbell-Thompson

Abstract

Immunohistochemistry (IHC) is a powerful immunology-based method that is used to study the location of proteins in cells and tissues. There have been numerous advancements in IHC technology that continually increase the sensitivity and specificity through which this method can be used to generate new discoveries. Similarly, Alpha-1 Antitrypsin (AAT) IHC can be used to study AAT protein expression within the human liver or exogenous AAT that is delivered through gene therapy. Here, we describe a highly sensitive method to detect the AAT antigen in formalin-fixed paraffin-embedded human or mouse tissues.

Key words Alpha-1 antitrypsin, Alpha-1 antitrypsin deficiency, Protein localization, Immunohistochemistry, Immunoperoxidase

1 Introduction

Alpha-1 antitrypsin deficiency is a hereditary disorder that was discovered in 1963, which occurs in 1 out of 2500–5000 newborns in the United States and Western Europe [1, 2]. The incidence is highly dependent on Scandinavian descent. AAT is a broad-spectrum, anti-inflammatory, immunomodulatory, and tissue repair molecule [2]. AAT deficiency results in the premature onset of chronic obstructive pulmonary disease, liver cirrhosis, and other inflammatory diseases. Mutations in the gene result in protein folding errors, which aggregate to form insoluble intracellular polymer inclusions [3]. Mutations also cause less circulating AAT in lung tissue to be available to neutralize elastase activity, leading to uncontrolled proteolytic attack and early onset pan-acinar emphysema [4, 5].

AAT IHC staining can be performed in order to assess mutated AAT accumulation levels in liver biopsies. Formalin-fixed paraffin-embedded tissue provides excellent tissue morphology and can be used for the detection of AAT in various tissues. In 1968, the introduction of peroxidase labeled secondary antibodies paved the

Florie Borel and Christian Mueller (eds.), *Alpha-1 Antitrypsin Deficiency: Methods and Protocols*, Methods in Molecular Biology, vol. 1639, DOI 10.1007/978-1-4939-7163-3_13, © Springer Science+Business Media LLC 2017

way for modern AAT IHC [6]. For the first antibody for AAT staining, the polymer-specific antibodies ATZ11 and 2C1 were developed in order to facilitate mechanism studies of mutated AAT polymer formation [7, 8]. Furthermore, with the development of polymer-based IHC, use of a polymer-horseradish peroxidase secondary antibody now provides significantly increased sensitivity for AAT IHC [9].

2 Materials

1. Formalin-fixed, paraffin-embedded, 4- to 5-μm-thick tissue sections on poly-L-lysine-coated or positively charged slides (e.g., Superfrost Plus slides, Fisher Scientific).

2. 60 °C oven.

3. Timer.

4. Xylenes.

5. Ethanols: 70%, 80%, 95%, 100% (v/v).

6. Tris-buffered saline (TBS), pH 7.6: Tris, NaCl, deionized water; 1 M HCl.

7. TBS-Tween: TBS, 0.05% (v/v) Tween 20.

8. 37 °C oven.

9. Digest-All™ Trypsin (Invitrogen): trypsin concentrate, diluent.

10. Sakura staining racks.

11. Tissue-Tek staining dishes.

12. 3% Hydrogen Peroxide in Methanol (v/v).

13. Distilled water (dH$_2$O).

14. Moist chamber (Evergreen Scientific).

15. Serum blocker (Sniper, Biocare Medical).

16. Antibody diluent (Vector Labs).

17. Primary antibody: AAT antibody (rabbit anti-human) at appropriate dilution (Fitzgerald Industries International).

18. Secondary antibody: Goat anti-rabbit HRP Polymer (Biocare Medical).

19. DAB (3,3 diaminobenzidene) (Vector Labs).

20. CAT Hematoxylin (Biocare Medical).

21. Cytoseal mounting medium (VWR Scientific).

22. Coverslips.

23. Brightfield microscope.

3 Methods

Carry out all procedures at room temperature in a humidified chamber unless otherwise specified.

1. Prepare the TBS as described prior to conducting this procedure:

   Weigh out 6.05 g of Tris, and 8.76 g NaCl. Dissolve completely in 800 ml of deionized water. Adjust the pH to 7.6 with 1 M HCl. Bring up to 1 l final volume with deionized water.

2. Dry the tissue sections on slides at 60 °C degrees for 1 h in the oven (*see* **Note 1**).

3. Deparaffinize and rehydrate the sections by placing the slides in Sakura staining racks and immersing them in Tissue-Tek staining dishes containing the following solutions (*see* **Note 2**):

   Two times in xylenes for 5 min each.

   Two times in 100% ethanol for 2 min each.

   Two times in 95% ethanol for 2 min each.

   One time in 70% ethanol for 1 min.

   Rinse the slides in running tap water for 1 min.

4. Shake off the excess water and place the slides in a staining dish containing 3% hydrogen peroxide in Methanol for 10 min in order to quench endogenous peroxidase activity (*see* **Note 3**).

5. Rinse the slides with distilled water for 5 min.

6. Shake off the excess water and add 300 μl of Digest-All Trypsin solution to cover the sections, place the slides in a moist chamber, and incubate them for 5 min in a 37 °C oven (*see* **Note 4**).

7. Rinse off the Trypsin solution with dH$_2$O for 1 min.

8. Place the slides in TBS-Tween for 5 min.

9. Drain off the solution and immediately add 300 μl of serum block (Sniper) to cover the sections and incubate them for 15 min in a moist chamber (*see* **Note 5**).

10. Rinse off the serum block in a staining dish of TBS-Tween.

11. Take the slides out of the TBS-Tween and drain off the solution. Immediately add 300 μl of anti-AAT to cover the sections and incubate the slides for 60 min in a moist chamber.

12. Wash the slides two times for 5 min each in TBS-Tween.

13. Shake off the TBS-Tween, apply 300 μl of the secondary antibody (goat anti- rabbit HRP polymer), and incubate the slides for 30 min.

14. Wash the slides two times for 5 min each in TBS-Tween.

15. Develop for 2–5 min in a DAB substrate solution (*see* **Note 6**). Terminate the reaction by rinsing the slides with dH$_2$O.

16. Counterstain with CAT hematoxylin for 30 s–1 min.
17. Rinse the slides with running tap water for 5 min.
18. Place the slides in TBS for 1 min (*see* **Note 7**).
19. Rinse the slides again with running tap water for 2 min.
20. Dehydrate the slides:

    One time in 80% ethanol for 1 min.

    Two times in 95% ethanol for 1 min.

    Two times in 100% ethanol for 1 min.

    Three times in xylenes for 1 min.
21. Coverslip with Cytoseal mounting medium.
22. Observe the IHC result under the brightfield microscope (Fig. 1).

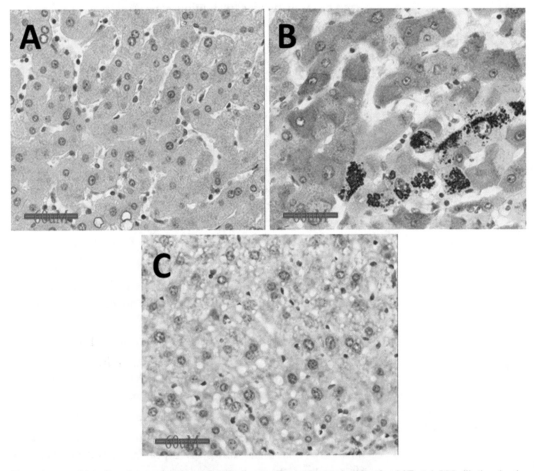

Fig. 1 Immunohistochemistry using formalin-fixed, paraffin-embedded slides for AAT at 1:800 dilution (under 40× lens). Diffuse cytoplasmic staining was observed in normal human hepatocytes (**a**) while a patient with AAT deficiency showed multiple positive intracellular globules (**b**). Mouse liver was used as negative control to demonstrate specificity of the primary antibody for the human protein (**c**)

4   Notes

1. Heating helps tissue sections adhere well to slides. Do not bake slides for more than 2 h.

2. Do not let sections dry once they are deparaffinized. Doing so would cause increased background.

3. Do not incubate the slides in 3% H_2O_2/Methanol for more than 10 min to avoid damaging protein antigenicity.

4. Preheat 300 µl of digestion buffer in the 37 °C oven and add 100 µl of Digest-All Trypsin immediately before applying it to the sections.

5. Optional: these steps can be carried out with a DAKO Autostainer.

6. CAUTION: DAB is a carcinogen and must be used with great care. Always wear gloves when handling it. Collect and dispose of waste according to local regulations.

7. TBS turns the red brown color of hematoxylin to blue, thereby increasing the contrast between DAB (brown) and hematoxylin (blue).

Acknowledgment

This work was supported by a grant from 5P01DK058327 (Dr. Barry Byrne, PI).

References

1. Wood AM, Stockley RA (2007) Alpha one antitrypsin deficiency: from gene to treatment. Respiration 74:481–492

2. de Serres F, Blanco I (2014) Role of alpha-1 antitrypsin in human health and disease. J Intern Med 276:311–335

3. Lomas DA, Evans DL, Finch JT, Carrell RW (1992) The mechanism of Z alpha 1-antitrypsin accumulation in the liver. Nature 357:605–607

4. Teckman JH, Mangalat N (2015) Alpha-1 antitrypsin and liver disease: mechanisms of injury and novel interventions. Expert Rev Gastroenterol Hepatol 9:261–268

5. Brebner JA, Stockley RA (2013) Recent advances in α-1-antitrypsin deficiency-related lung disease. Expert Rev Respir Med 7:213–229. quiz 230

6. Callea F, Brisigotti M, Faa G, Lucini L, Eriksson S (1991) Identification of PiZ gene products in liver tissue by a monoclonal antibody specific for the Z mutant of alpha 1-antitrypsin. J Hepatol 12:372–376

7. Janciauskiene S, Dominaitiene R, Sternby NH, Piitulainen E, Eriksson S (2002) Detection of circulating and endothelial cell polymers of Z and wild type alpha 1-antitrypsin by a monoclonal antibody. J Biol Chem 277:26540–26546

8. Miranda E, Pérez J, Ekeowa UI, Hadzic N, Kalsheker N, Gooptu B, Portmann B, Belorgey D, Hill M, Chambers S, Teckman J, Alexander GJ, Marciniak SJ, Lomas DA (2010) A novel monoclonal antibody to characterize pathogenic polymers in liver disease associated with alpha1-antitrypsin deficiency. Hepatology 52:1078–1088

9. Mueller C, Chulay JD, Trapnell BC, Humphries M, Carey B, Sandhaus RA, McElvaney NG, Messina L, Tang Q, Rouhani FN, Campbell-Thompson M, AD F, Yachnis A, Knop DR, Ye GJ, Brantly M, Calcedo R, Somanathan S, Richman LP, Vonderheide RH, Hulme MA, Brusko TM, Wilson JM, Flotte TR (2013) Human Treg responses allow sustained recombinant adeno-associated virus-mediated transgene expression. J Clin Invest 123:5310–5318

Chapter 14

Periodic Acid-Schiff Staining with Diastase

Dongtao A. Fu and Martha Campbell-Thompson

Abstract

Periodic Acid-Schiff (PAS) with diastase (PAS-D) refers to the use of the PAS stain in combination with diastase, which is an enzyme that digests the glycogen. The purpose of using the PAS-D procedure is to differentiate glycogen from other PAS-positive elements in tissue samples. The PAS-D method is also used for periportal liver staining of AAT polymer inclusions that are seen in alpha-1 antitrypsin deficiency disease. Here, we describe the procedure of PAS-D staining in formalin-fixed, paraffin-embedded human liver tissues.

Key words Special stain, Periodic Acid-Schiff, Periodic Acid-Schiff reaction-diastase, Alpha-1 antitrypsin deficiency, Human liver

1   Introduction

The Periodic Acid-Schiff (PAS) reaction and Periodic Acid-Schiff reaction with diastase (PAS-D) are two special staining procedures that are commonly performed in a histology laboratory. PAS-D is a very sensitive histochemical method for the examination of glycogen. Diastase, which is also referred to as α-amylase, acts on glycogen to depolymerize it into smaller sugar units (maltose and glucose) that are washed out of the section [1]. The intensities of the two stains (PAS and PAS-D) are associated with differential glycogen concentrations and can be used to compare the glycogen level in samples [2].

Alpha-1 antitrypsin deficiency is an inherited metabolic disorder in which mutations in the coding sequence of the Glu342Lys protein prevent its export from a hepatocyte [3]. In 1975, the first report of a histopathologic examination of liver specimens from alpha-1 antitrypsin deficiency patients demonstrated the classic intracellular hyaline globules that stain positive with PAS-D staining [4]. Their location was highest in the periportal region (Type 1 globular) [5]. The globules are large and most commonly seen in the PiZZ genotype but can also be seen in the PiMZ genotype.

Florie Borel and Christian Mueller (eds.), *Alpha-1 Antitrypsin Deficiency: Methods and Protocols*, Methods in Molecular Biology, vol. 1639, DOI 10.1007/978-1-4939-7163-3_14, © Springer Science+Business Media LLC 2017

In patients with manifestations of liver disease, a liver biopsy provides a sample for PAS-D staining that is valuable for studying the progression of this condition.

2   Materials

All materials should be prepared at room temperature and stored at 4 °C unless otherwise indicated.

1. Formalin-Fixed, paraffin-embedded, 4- to 5-µm-thick tissue sections on poly-L-lysine-coated, charged slides (e.g., Plus slides, Fisher Scientific).

2. 60 °C oven.

3. Sakura staining racks.

4. Tissue-Tek staining dishes.

5. Xylenes.

6. Ethanol: 70%, 80%, 95%, 100% (v/v).

7. Oxoid™ PBS Tablets (Thermo-Fisher Scientific).

8. Sodium Azide.

9. Alpha Amylase from Bacillus Subtillis: 1 g Alpha Amylase, 0.1 g Sodium Azide, 100 mL Distilled water, 1 Oxoid™ PBS tablet (*see* **Note 1**).

10. 0.5% Periodic acid: 1 g Periodic acid powder, 200 mL distilled water (*see* **Note 2**).

11. Schiff's reagent, store in 4 °C refrigerator.

12. Hematoxylin 7211.

13. Clarifier™ 1 (Richard Allan Scientific).

14. Bluing Reagent.

15. Mounting medium (Cytoseal).

16. Coverslips.

3   Methods

Carry out all the procedures at room temperature unless otherwise specified. Preheat the Schiff's reagent at room temperature for at least 30 min.

Prepare the following two solutions as described prior to conducting this procedure.

1. Alpha Amylase from Bacillus Subtillis: Add 100 mL of Distilled water to a glass beaker. Weigh 1 g of Alpha Amylase and 0.1 g of Sodium Azide and transfer both to the beaker. Add one

tablet of Oxoid™ PBS to the beaker. Mix well and store at 4 °C when not in use (*see* **Note 1**).

2. 0.5% Periodic acid: Add 200 mL of distilled water to a glass beaker. Weigh 1 g of Periodic powder and transfer it to the beaker. Mix well (*see* **Note 2**).

3. Dry the sections on slides at 60 °C for 15 min in the oven (*see* **Note 3**).

4. Deparaffinize and rehydrate the tissue by placing the slides in Sakura staining racks and submerging them in Tissue-Tek staining dishes containing the following solutions:

   Two times in xylenes for 5 min.

   Two times in 100% ethanol for 2 min.

   Two times in 95% ethanol for 2 min.

   One time in 70% ethanol for 1 min.

   Rinse the slides in dH$_2$O for 1 min.

5. Place the slides in the amylase solution for 20 min.

6. Wash the slides well in dH$_2$O.

7. Place the slides in 0.5% periodic acid for 10 min.

8. Wash three times with dH$_2$O, changing the water each time.

9. Place the slides in Schiff's reagent for 15 min (*see* **Notes 4 and 5**).

10. Wash in running lukewarm tap water for 5–10 min in order for the tissue to develop a pink color (*see* **Note 6**).

11. Counterstain with hematoxylin 7211 for 1 min.

12. Wash three times with dH$_2$O, changing the water each time.

13. Place the slides in Clarifier™ 1 for 30 s.

14. Rinse the slides in dH$_2$O for 30 s.

15. Place the slides in bluing reagent for 30 s in order to differentiate the blue color.

16. Rinse the slides in dH$_2$O for 1 min.

17. Dehydrate the slides in the following solutions:

    One time in 80% ethanol for 1 min.

    Two times in 95% ethanol for 1 min.

    Two times in 100% ethanol for 1 min.

    Three times in xylenes for 1 min.

18. Coverslip the slides using a synthetic mounting medium (e.g., Cytoseal).

19. Observe the PAS-D result under the brightfield microscope (Fig. 1).

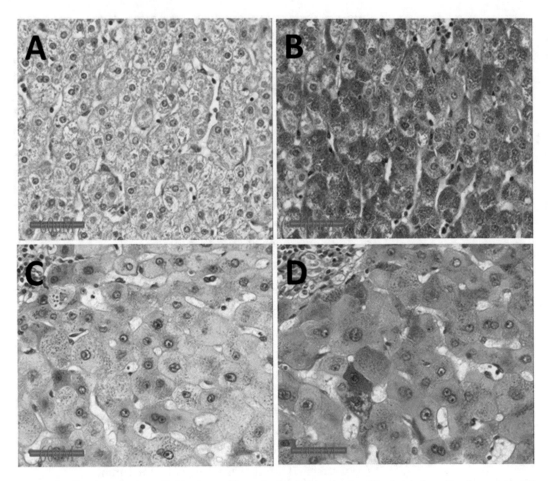

Fig. 1 PAS-D stain using formalin-fixed, paraffin-embedded slides (under 40× lens). A section from a normal human liver stained with PAS after diastase digestion (**a**) and a duplicate of the section stained with a PAS stain (**b**). A section from the liver sample of a patient with AAT deficiency stained with PAS after diastase digestion (**c**). A duplicate of the section stained with a PAS stain (**d**)

4   Notes

1. The Amylase solution can be stored up to 1 month at 4 °C.

2. Make fresh 0.5% periodic acid each time this procedure is performed.

3. Use PAS-positive tissues (kidney and liver) for quality control purposes.

4. In order to test the Schiff reagent, place 10 mL of 37% formalin into a watch glass and add a few drops of the Schiff reagent to it. Good Schiff reagent will rapidly turn a red-purple color. Deteriorating Schiff's reagent will display a delayed reaction and the color produced will be a deep blue-purple [6].

5. Keep the container with Schiff's reagent under a fume hood to contain the fumes.

6. Washing helps remove any excess reagent from the section and promotes the development of the rich pink color.

Acknowledgment

This work was supported by a grant from 5P01DK058327 (Dr. Barry Byrne, PI).

Reference

1. Carson FL (1997) Histotechnology: a self-instructional text, 2nd edn. ASCP Press, Chicago, pp 115–116

2. Web information: https://en.wikipedia.org/wiki/PAS_diastase_stain

3. Fairbanks KD, Tavill AS (2008) Liver disease in alpha 1-antitrypsin deficiency: a review. Am J Gastroenterol 103:2136–2141

4. Eriksson S, Larsson C (1975) Purification and partial characterization of PAS-positive inclusion bodies from the liver in alpha1-antitrypsin deficiency. N Engl J Med 292:176–180

5. Qizilbash A, Young-Pong O (1983) Alpha 1 antitrypsin liver disease differential diagnosis of PAS-positive, diastase-resistant globules in liver cells. Am J Clin Pathol 79(6):697–702

6. IHCWORLD   http://www.ihcworld.com/_protocols/special_stains/pas_diastase_ellis.htm

Chapter 15

Protocol for Directed Differentiation of Human Induced Pluripotent Stem Cells (iPSCs) to a Hepatic Lineage

Joseph E. Kaserman and Andrew A. Wilson

Abstract

Directed differentiation is a powerful cell culture technique where developmental pathways are applied to a pluripotent progenitor in order to generate specific terminally differentiated cell populations. Here, we describe a serum-free protocol using growth factors in defined concentrations to derive iPSC-hepatic cells starting from both feeder and feeder-free conditions. The generated iPSC-hepatic cells are developmentally similar to fetal stage hepatocytes, and when generated from patients with genetic mutations such as alpha-1 antitrypsin deficiency recapitulate pathologic changes associated with clinical disease, such as protein misfolding, intracellular retention of misfolded proteins, and elevated levels of ER stress.

Key words Induced pluripotent stem cell (iPSC), Directed differentiation, Definitive endoderm, iPSC-hepatic cell

1 Introduction

The ability to maintain embryonic stem cells (ESCs) in an in-vitro culture system has been well described for over 30 years [1]. Their essential characteristics, including the ability to self-renew and differentiate to all three germ layers (ectoderm, endoderm, and mesoderm), resulted in significant interest in their potential therapeutic applications [2]. However, ethical concerns about the destruction of human embryos necessary to generate each genetically distinct ESC line limited progress within the field. In 2006, the field of stem cell biology was dramatically altered by the seminal discovery of Takahashi and Yamanaka that by overexpressing four transcription factors (Oct4, Klf4, Sox2, and cMyc) it was possible to reprogram a terminally differentiated somatic cell back to a pluripotent state [3]. These cells were given the name induced pluripotent stem cells (iPSCs), and as they are created from adult cells such as fibroblasts and monocytes they do not carry the stigma of their ESC counterparts [4, 5]. Additionally, it was now possible to

Florie Borel and Christian Mueller (eds.), *Alpha-1 Antitrypsin Deficiency: Methods and Protocols*, Methods in Molecular Biology, vol. 1639, DOI 10.1007/978-1-4939-7163-3_15, © Springer Science+Business Media LLC 2017

generate iPSCs from patients with specific genetic and phenotypic characteristics [6].

The introduction of iPSCs has been accompanied by advancements in the ability to derive specific cell lineages from pluripotent progenitors. By providing key growth factors at specific time points, mimicking signals and developmental pathways from the embryo, it is possible to pattern these pluripotent cells into a specific cell fate through a process known as directed differentiation [7]. Multiple differentiated cell types have been successfully derived from iPSCs; however, directed differentiation protocols have been characterized by wide variation in efficiency and maturation potential [8–10]. The ability to derive iPSC-hepatic cells has become well established and represents one of the more efficient protocols as it is possible to achieve >90% iPSC-hepatic cells as marked by AAT+/FOXA1+ cells [10]. To derive iPSC-hepatic cells, pluripotent progenitors are first exposed to the nodal analogue Activin A to derive definitive endoderm as defined by the cellular phenotype ckit+/CXCR4+/FOXA2+/brachyury- [11]. Definitive endoderm is then specified to the hepatic lineage through exposure to high-dose BMP-4 and FGF-2 [12]. Then through treatment with additional growth factors including HGF, gamma secretase inhibitor, and Oncostatin M, the cells undergo maturation [10]. Importantly, after 25 days these cells demonstrate key functional characteristics of fetal hepatocytes and, when derived from patients with genetic disorders (such as alpha-1 antitrypsin deficiency), recapitulate critical components of disease pathogenesis (Fig. 1) [10, 13]. In contrast to primary hepatocytes, which are limited in their ability to proliferate and lose their phenotype in culture, iPSC-hepatic cells provide an unlimited supply of human derived hepatic cells. Additionally, the ability to generate iPSCs from patients with known clinical phenotypes provides the unique opportunity to interrogate specific genetic contributions to liver disease pathogenesis. Finally, there is great interest in whether iPSC-hepatic cells can be used in personalized medicine applications such as drug testing [10].

2   Materials

2.1 Human Cell Culture Components

1. Matrigel. Store at −80 °C (BD Biosciences).

2. Gelatin 0.1% (EMD Millipore).

3. mTeSR1 medium: Maintenance culture medium for hiPSCs, prepare by combining supplied 5× supplement with base medium. Store at 4 °C (Stem Cell Technologies).

4. Dulbecco's Modified Eagle Medium (DMEM): high glucose, L-Glutamine, sodium pyruvate. Store at 4 °C (Thermo Fisher Scientific).

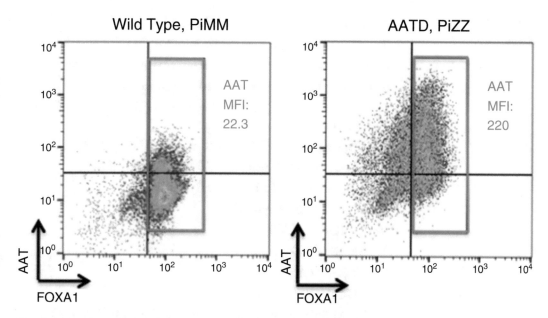

Fig. 1 Intracellular AAT Accumulation in PiZZ Alpha-1 Antitrypsin Deficiency iPSC-Hepatic Cells. Flow cytometry of fixed, permeabilized iPSC-hepatic cells using anti-AAT and anti-FOXA1 antibodies demonstrates a significantly increased level of intracellular AAT protein in PiZZ iPSC-hepatic cells compared to wild-type PiMM cells. This is quantified by the main fluorescence intensity (MFI) which approximates the amount of stained intracellular AAT protein

5. Dulbecco's Modified Eagle Medium: Nutrient Mixture F-12 (DMEM/F12). Store at 4 °C (Thermo Fisher Scientific).

6. Iscove's Modified Dulbecco's Media (IMDM). Store at 4 °C (Thermo Fisher Scientific).

7. Dulbecco's Phosphate-Buffered Saline (DPBS): No calcium, no magnesium. Store at 4 °C (Thermo Fisher Scientific).

8. Gentle Cell Dissociation Reagent (Stem Cell Technologies).

9. Collagenase Type IV. Store at −4 °C (Life Technologies).

10. Primocin. Store at −20 °C (Thermo Fisher Scientific).

11. Rho-associated protein kinase (ROCK) inhibitor Y-27632. Store at −80 °C until thawed and then store at 4 °C (Tocris).

12. Defined Fetal Bovine Serum (FBS) heat inactivated. Store at −20 °C (Thermo Fisher Scientific).

13. Knockout Serum. Store at −20 °C (Thermo Fisher Scientific).

14. Non-Essential Amino Acid (NEAA). Store at 4 °C (Life Technologies).

15. Glutamax. Store at 4 °C (Life Technologies).

16. 2-Mercaptoethanol. Store at 4 °C (Life Technologies).

2.2 Differentiation Components

1. Ham's F-12. Store at 4 °C (Cellgro).
2. L-Glutamine- 200 mM. Store at −20 °C (Invitrogen).
3. B27 without Retinoic Acid. Store at −20 °C (Gibco).
4. N2 Supplement. Store at −20 °C (Gibco).
5. Bovine Albumin Fraction V (BSA)—7.5% solution. Store at 4 °C (Invitrogen).
6. 1-Thioglycerol (MTG). Store at −20 °C (Sigma-Aldrich).
7. L-Ascorbic Acid. Store at −20 °C (Sigma-Aldrich).
8. rhBMP4 (*see* **Note 1**) (R&D Systems).
9. rhFGF-2 (R&D Systems).
10. rhVEGF (R&D Systems).
11. rhEGF (R&D Systems).
12. rhTGFα (R&D Systems).
13. rhHGF (R&D Systems).
14. STEMdiff Definitive Endoderm Kit (Stem Cell Technologies).
15. Dexamethasone. Store at −20 °C (Sigma-Aldrich).
16. Oncostatin-M. Store at −20 °C (Invitrogen).
17. Vitamin-K. Store at −20 °C (Invitrogen).
18. Gamma Secretase Inhibitor (gSI). Store at −20 °C (EMD Millipore).

2.3 Antibodies

1. APC anti-human CD117 (c-kit) Antibody. Store at 4 °C (BioLegend).
2. Mouse IgG1, APC (c-kit isotype). Store at 4 °C (Invitrogen).
3. CXCR4/CD184 Antibody, RPE conjugate. Store at 4 °C (Invitrogen).
4. Mouse IgG2a, (R-PE) (CXCR4 isotype). Store at 4 °C (Invitrogen).

3 Methods

All cell culture work should be carried out in a biosafety cabinet with laminar flow using a proper sterile technique.

3.1 Hepatic-Directed Differentiation from Feeder-Free Starting Conditions

1. Human iPSC lines should be maintained in culture for at least 1 week prior to using them for differentiation (*see* **Note 2**). For maintenance, cells should be grown on matrigel-coated tissue culture plates in mTeSR-1 media supplemented with primocin 50 μg/mL (may use pen/strep as well).
2. Cells should be grown until confluent and ready to pass (*see* **Note 3**).

3. Prepare the necessary number of 6-well plates by coating with matrigel. 1 mg of matrigel should be diluted in 12 mL of cold DMEM/F12. Use 1 mL of matrigel-DMEM/F12 mixture to coat 1 well of a 6-well plate. Plates should sit for 1 h at room temperature or 30 min at 37 °C.

4. Aspirate the matrigel and wash once with 2 mL of DMEM/F12. Add 2 mL of mTeSR-1 supplemented with 10 μM ROCK inhibitor.

5. Aspirate media from cells to be passaged and wash with 2 mL PBS. Add 1 mL of Gentle Cell Disassociation (GCD) Reagent to each well and incubate at 37 °C for 6 min.

6. Aspirate and add 1 mL of DMEM/F12 and gently wash cells with a P1000 pipette. Transfer to a 15 mL conical (if needed can perform an additional wash step with 1 mL of media to harvest persistently adherent cells).

7. Centrifuge at $300 \times g$ and 4 °C for 5 min.

8. Aspirate the supernatant and resuspend in 1 mL of mTeSR-1 supplemented with 10 μM ROCK inhibitor. Gently pipette ten times until a single cell suspension is achieved. Use 10 μL of this suspension to then perform a cell count.

9. Plate 1×10^6 cells/well of a 6-well plate (*see* **Note 4**) onto the prepared matrigel plates. The cells should then be moved into a 5% O_2 hypoxic incubator for the remainder of the protocol. This is considered day 0 of the differentiation protocol.

3.2 Hepatic-Directed Differentiation Starting from Feeders

1. For maintenance, iPSCs are grown on MEF-feeder plates in hiPS media (*see* **Notes 5** and **6**).

2. Cells should be grown until ~70% confluent and ready to pass.

3. Prepare the necessary number of 6-well plates by coating with matrigel, as described above in the feeder-free protocol. Note unlike the feeder-free protocol, cells in this protocol are passed as clumps rather than counted in single-cell suspension. Typically, cells are passed at a split ratio ranging from 1:1 to 1:3.

4. Aspirate hiPS media from cells to be passed and wash cells with 2 mL of DMEM/F12. Add 1 mL of collagenase solution (10 mg/mL) and incubate for 2 min at 37 °C.

5. Aspirate collagenase and wash with 2 mL DMEM/F12. Add 1 mL of hiPS media with 10 μM ROCK inhibitor. Wash and gently scrape 5–10× with a P1000 to detach cells. Note cells will still be in clumps. Transfer to the matrigel-coated well(s) containing an additional 1 mL of hiPS media with 10 μM ROCK inhibitor.

6. Refeed daily with hiPS media for 24–72 h until well is at least 80% confluent. At this point can begin with day 1 of the definitive endoderm step and cells should be moved to a 5% O_2 hypoxic incubator for the remainder of the protocol.

3.3 Definitive Endoderm Days 1–5

1. Prepare day 1 media by mixing 980 μL basal medium with 10 μL of supplement A and 10 μL of supplement B per mL of media using the STEMdiff definitive endoderm kit. We will need to prepare 2 mL of media for each well and supplement with 50 μg/mL of primocin.

2. Aspirate media and wash with 2 mL of DMEM/F12. Feed with day 1 media for 24 h.

3. Prepare days 2–4 media by mixing 990 μL basal medium with 10 μL of supplement B per mL of media and primocin 50 μg/mL. We will need to prepare sufficient volume to refeed with 2 mL of media per well each day.

4. Aspirate media and wash with DMEM/F12. Feed with 2 mL of days 2–4 media and then refeed every 24 h.

5. Starting at day 5 all media will use the same CSFDM base and should prepare enough for the remainder of the differentiation (*see* **Note 7**). By day 5 the wells are usually completely confluent (Fig. 2).

6. Prepare days 5–6 media by adding BMP4 10 ng/mL, FGF2 10 ng/mL, and VEGF 10 ng/mL to CSFDM.

7. Prepare new matrigel plates with day 5–6 media supplemented with 10 μM of ROCK inhibitor.

8. Aspirate old media and wash with 2 mL of DMEM/F12. Add 1 mL of GCD and incubate at 37 °C for 2 min. Note: save 1 well or a portion of a well for FACS analysis to quantify efficiency of differentiation to definitive endoderm (see the protocol below).

9. Aspirate GCD and add 1 mL of day 5 media.

10. Gently wash cells by pipetting 5–10 times with P1000 pipette. A confluent well can be passed 1:4 into 4 wells of a new 6-well plate (or 8 wells of a 12-well plate, etc.) (*see* **Note 8**).

| Undifferentiated iPSCs | Definitive Endoderm Day 5 | Hepatic Progenitor Day 13 | iPSC-hepatic cell Day 25 |

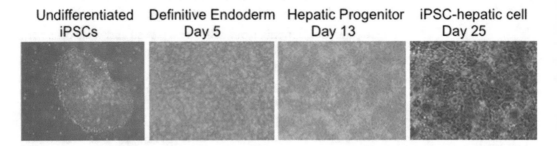

Fig. 2 Morphological Changes at Key Time Points during iPSC-hepatic Cell Directed Differentiation. The addition of the modal analogue Activin A to undifferentiated cells helps drive them to definitive endoderm. By day 5, cells no longer form tight distinct colonies but instead have informed a confluent monolayer. Following passaging and treatment with high dose BMP4, HGF, and dexamethasone, the cells begin to assume features typical of hepatocytes, and with further maturation by day 25 have the classic polygonal morphology

3.4 Day 5 FACS Analysis

1. Collect cells as above (can use either a whole well or a portion of a well if cell numbers are limited) as well as an undifferentiated control well.

2. If using less than an entire well bring volume up to 1 mL. Pipette 10–15 times and use 10 μL for cell count.

3. Centrifuge cells at $300 \times g$ for 5 min at 4 °C.

4. Aspirate the supernatant and resuspend in 250 μL of PBS+ (49 mL DPBS and 1 mL FBS) (*see* **Note 9**).

5. Aliquot 50 μL into each eppendorf on ice.

6. Add primary antibodies to each tube: 2 μL of each isotype/1×10^6 cells, 5 μL of ckit/1×10^6 cells, 5 μL of CXCR4/1×10^6 cells.

7. Vortex briefly and incubate on ice covered with foil for 30 min.

8. Add 1 mL of PBS+, vortex and centrifuge at $300 \times g$ and 4 °C for 5 min.

9. Aspirate and resuspend in 400 μL of FACS buffer pipetting 10–15× to resuspend. Filter using 40 μm filter into FACS tube and keep on ice covered with foil until FACS is performed (*see* Fig. 3 and **Note 10**).

3.5 iPSC-Hepatic Cells Days 6–25

1. Aspirate media and refeed with 2 mL of day 5–6 media.

2. Prepare day 7–12 media by adding BMP4 50 ng/mL, FGF2 10 ng/mL, VEGF 10 ng/mL, EGF 10 ng/mL, TGFα

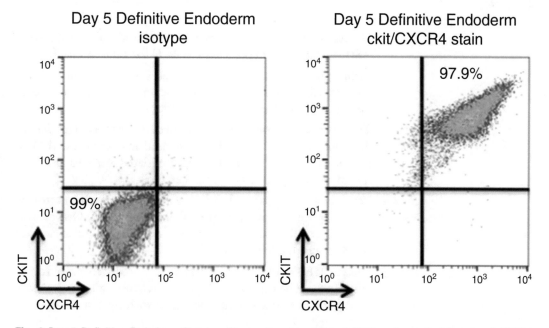

Fig. 3 Day 5 Definitive Endoderm Staining. Flow cytometry of day 5 iPSCs using anti-ckit and anti-CXCR4 antibodies demonstrates highly efficient induction of definitive endoderm

20 ng/mL, HGF 100 ng/mL, and dexamethasone 0.1 μM in CSFDM base media (*see* **Note 11**).

3. Note before switching between days 7–12, 13–18, and 19–25 medias wash wells once with 2 mL of IMDM to remove any residual growth factors prior to adding the new media. Feed with 2 mL of day 7–12 media. From this point in the protocol cells only need to be refed every 48 h with fresh media.

4. Prepare day 13–18 media by adding FGF2 10 ng/mL, VEGF 10 ng/mL, EGF 10 ng/mL, HGF 100 ng/mL, oncostatin M 20 ng/mL, vitamin K 6 μg/mL, dexamethasone 0.1 μM, gSI 1.5 μM (*see* **Note 12**), and DMSO 1% in CSFDM base (*see* **Note 13**). Refeed with 2 mL every 48 h.

5. Prepare day 19–25 by adding HGF 100 ng/mL, oncostatin M 20 ng/mL, vitamin K 6 μM, and dexamethasone 0.1 μM in CSFDM base. Refeed with 2 mL every 48 h.

6. By day 25 the wells should appear completely confluent with the overwhelming majority of cells having the classic hepatocyte polygonal morphology (Fig. 2). If desired supernatants can be collected and analyzed for protein secretion. By this time-point the cells have matured sufficiently that they can be harvested for intracellular FACS analysis (Fig. 1), protein extraction for western blots, RNA extraction (*see* **Note 14**), or immunostaining.

4 Notes

1. Unless otherwise specified growth factors should be stored at −80 °C and once thawed kept at 4 °C for up to a month.

2. Although this protocol has been used successfully on multiple distinct wild-type and disease-specific human-iPSC lines, we do observe line-to-line variability in efficiency.

3. The number of confluent wells needed is experiment dependent and reflects the number of conditions being tested. Also be sure to save a well of undifferentiated cells to use as a negative control for day 5 FACS analysis.

4. We have found plating 1–1.5×10^6 cells per well in a 6-well format appears to provide optimal density at day 5. However, as few as 7×10^5 and as many as 2×10^6 cells per well can still provide acceptable results.

5. Feeder MEF's need to be mitotically inactivated prior to use with iPSCs. They are available commercially or alternatively can be expanded in house and inactivated with mitomycin-c.

6. hiPS media: 200 mL DMEM/F12, 50 mL knockout serum, 2.5 mL Non-Essential Amino Acids, 2.5 mL glutamax, 500 μL 2-mercaptoethanol, primocin 50 μg/mL, and FGF2 5 ng/mL.

7. CSFDM media: IMDM 375 mL, Ham's F12 125 mL, B27 without retinoic acid 5 mL, N2 supplement 2.5 mL, 7.5% BSA 3.3 mL, L-glutamine 5 mL, ascorbic acid 5 mL, Diluted MTG (13 μL MTG diluted in 1 mL IMDM) 1.5 mL, and primocin 50 μg/mL. After combining the components cover with foil as the media is light sensitive.

8. If starting from feeders split no more than 1:3 as there is increased cell death when compared to cells that have already been adapted to feeder-free conditions.

9. Can adjust the volume of PBS+ used depending on the number of samples needed. For day 5 definitive endoderm cells will need five conditions: an unstained control, isotypes, single ckit, single CXCR4, and double stain ckit/CXCR4. For undifferentiated controls only need two conditions: unstained control, double stain ckit/CXCR4.

10. We routinely see 90+% of cells stain double positive for ckit/CXCR4 (Fig. 3). If endoderm induction is poor (<80%) we would recommend stopping at this point and restarting until a more efficient endodermal induction is achieved. Additionally, it is important that if using multiple different iPSC lines they have similar levels of definitive endoderm or downstream data interpretation could be confounded by the differences in efficiency.

11. Once growth factors are added to the CSFDM base media it lasts not more than 2 weeks and ideally should be used within 7 days.

12. gSI is light sensitive and needs to be protected even during storage at −20 °C.

13. By day 13 we expect that cells will have started to assume the morphological characteristics of hepatocytes (Fig. 2). Additionally, cells have begun to have visible prominent intracellular inclusions by light microscopy.

14. For RNA extractions using RLT+ buffer provides low but sufficient yields for qPCR analysis from one well of a 6-well plate. However, if using small well sizes or only a fraction of a well we recommend performing a QIAzol extraction to ensure adequate RNA yields.

Acknowledgments

This work was supported by NIH/NIDDK grant R01DK101501 and the Alpha-1 Foundation.

References

1. Evans MJ, Kaufman MH (1981) Establishment in culture of pluripotent cells from mouse embryos. Nature 292:154–156

2. Thomson JA, Itskovitz-Eldor J, Shapiro SS et al (1998) Embryonic stem cell lines derived from human blastocysts. Science:1145–1147

3. Takahashi K, Yamanaka S (2006) Induction of pluripotent stem cells from mouse embryonic and adult fibroblast cultures by defined factors. Cell 126:663–676

4. Takahashi K, Tanabe K, Ohnuki M et al (2007) Induction of pluripotent stem cells from adult human fibroblasts by defined factors. Cell 131:861–872

5. Sommer AG, Rozelle SS, Sullivan S et al (2012) Generation of human induced pluripotent stem cells from peripheral blood using the STEMCCA lentiviral vector. J Vis Exp 68: e4327

6. Somers A, Jean JC, Omari A et al (2010) Generation of transgene-free lung disease-specific human iPS cells using a single excisable lentiviral stem cell cassette. Stem Cells 28:1728–1740

7. Irion S, Nostro MC, Kattman SJ et al (2008) Directed differentiation of pluripotent stem cells: from developmental biology to therapeutic applications. Cold Spring Harb Symp Quant Biol 73:101–110

8. Karumbayaram S, Novitch BG, Patterson M et al (2009) Directed differentiation of human-induced pluripotent stem cells generates active motor neurons. Stem Cells 27:806–811

9. Shiba Y, Fernandes S, Zhu WZ et al (2012) Human ES-cell derived cardiomyoctes electrically couple and suppress arrhythmias in injured hearts. Nature 489:322–325

10. Wilson AA, Ying L, Liesa M et al (2015) Emergence of a stage-dependent human liver disease signature with directed differentiation of alpha-1 antitrypsin-deficient iPS cells. Stem Cell Reports 4:1–13

11. Kubo A, Shinozaki K, Shannon JM et al (2004) Development of definitive endoderm from embryonic stem cells in culture. Development 131:1651–1662

12. Gouon-Evans V, Boussemart L, Gadue P et al (2006) BMP-4 is required for hepatic specification of mouse embryonic stem cell-derived definitive endoderm. Nat Biotechnol 24:1402–1411

13. Tafaleng EN, Chakraborty S, Han B et al (2015) Induced pluripotent stem cells model personalized variations in liver disease resulting from α1-antitrypsin deficiency. Hepatology 62 (1):147–157

Chapter 16

Isolation of Kupffer Cells and Hepatocytes from a Single Mouse Liver

Marcela Aparicio-Vergara, Michaela Tencerova, Cecilia Morgantini, Emelie Barreby, and Myriam Aouadi

Abstract

Liver perfusion is a common technique used to isolate parenchymal and non-parenchymal liver cells for in vitro experiments. This method allows hepatic cells to be separated based on their size and weight, by centrifugation using a density gradient. To date, other methods allow the isolation of only one viable hepatic cellular fraction from a single mouse; either parenchymal (hepatocytes) or non-parenchymal cells (i.e., Kupffer cells or hepatic stellate cells). Here, we describe a method to isolate both hepatocytes and Kupffer cells from a single mouse liver, thereby providing the unique advantage of studying different liver cell types that have been isolated from the same organism.

Key words Liver, Perfusion, Kupffer cells, Hepatocytes, Cell isolation, Gradient centrifugation

1 Introduction

The liver plays a crucial role in metabolic and immune system functions [1–4]. This organ is comprised of two major cells types, parenchymal and non-parenchymal cells. The most abundant hepatic parenchymal cells are the hepatocytes, which constitute 92.5% of the liver [5]. Hepatocytes facilitate important metabolic functions in the body including the control of triglyceride levels as well as cholesterol biosynthesis and uptake [6, 7]. Moreover, hepatocytes participate in the regulation of carbohydrate metabolism as well as bile production in order to aid the digestion and uptake of lipids from the intestines [8].

The second major cell type of the liver, the non-parenchymal cells, contains 70% endothelial cells, 20% Kupffer cells (KCs), and 10% stellate cells in rodents [5, 9]. KCs, the resident macrophages in the liver, represent 80–90% of all tissue macrophages in the body of rodents and are located in the sinusoids, through which they come into contact with nutrients, microorganisms, and toxic agents

Florie Borel and Christian Mueller (eds.), *Alpha-1 Antitrypsin Deficiency: Methods and Protocols*, Methods in Molecular Biology, vol. 1639, DOI 10.1007/978-1-4939-7163-3_16, © Springer Science+Business Media LLC 2017

that are transported by the hepatic system [10]. These particular cells are responsible for the clearance of foreign material such as bacteria [11]. Upon recognition of a foreign substance, KCs undergo activation and release a variety of regulatory factors such as cytokines, chemokines, proteolytic enzymes, nitric oxide, and reactive oxygen species [12]. KCs that become activated as a result of liver injury due to steatohepatitis and metabolic complications are the major source of inflammatory mediators that ultimately lead to hepatic toxicity and the impairment of liver metabolism. Accordingly, KCs are thought to be the major source of hepatic inflammation [12–17].

The important role of hepatocytes and KCs in homeostasis has emphasized the need to develop techniques that can be used to isolate and study these cells. To date, multiple protocols involving liver perfusion, enzymatic digestion, and gradient centrifugation have been developed [18, 19]. However, most of these methods present several limitations, including: (1) the isolation of only one hepatic cellular fraction, (2) the requirement of expensive equipment, and (3) low reproducibility. Therefore, there is a need for an isolation technique that can be performed in a rapid, economical, and reproducible manner that will lead to functional, intact, and enriched hepatocyte and KC fractions from a single mouse liver.

Here, we describe a remarkably cost-effective and reproducible protocol for isolating hepatocytes and KCs from a single mouse liver. This method is based on hepatic perfusion with collagenase, followed by the separation of the hepatic cells according to their density using low-speed and gradient centrifugation.

2   Materials

2.1 Solutions for Liver Perfusion and Cell Isolation

The following solutions will need to be prepared and stored before the protocol is conducted. Prepare all solutions using ultrapure water and store all reagents at 4 °C unless otherwise specified.

2.2 Liver Perfusion Reagents

1. 1× HBSS buffer: 10× HBSS, ultrapure water.

2. HBSS-EGTA buffer (0.5 mM): 500 μL EGTA 0.5 M, 500 mL 1× HBSS.

3. HBSS-CaCl$_2$ buffer (1 mM): 500 μL of 1 M CaCl$_2$, 500 mL 1× HBSS (*see* **Note 1**).

4. Concentrated Collagenase: 1.6 g of collagenase from *Clostridium histolyticum*, 50 mL HBSS-CaCl$_2$ buffer.

5. Mouse anesthesia: 200 μL per mouse of 100 mg/kg of Ketamine, 10 mg/kg Xylazine, phosphate buffer solution (PBS).

2.3 Density Gradient for KC Isolation Using Percoll

1. 90% Percoll: 5 mL 10× PBS, 45 mL Percoll.
2. 25% Percoll: 30 mL 1× PBS, 10 mL of 90% Percoll.
3. 50% Percoll: 15 mL 1× PBS, 15 mL of 90% Percoll (*see* **Note 2**).
4. Percoll gradient: 20 mL 25% Percoll, 14.5 mL 50% Percoll (*see* **Note 3**).

2.4 Media for the Cultivation of Isolated Liver Cells

1. Adherence medium M199 for hepatocytes: 475 mL of M199 + Earles Salts, 5 mL of Penicillin/Streptomycin (P/S), 5 mL of BSA (10%), 10 mL of UltroSer™ G (Pall BioSepra) (*see* **Note 4**) or Fetal Bovine Serum, 5 μL (100 nM final) of Dexamethasone, 72.5 μL (100 nM final) of Insulin.
2. Basal (maintenance) medium for hepatocytes: 475 mL of M199 + Earles Salts, 5 μL (100 nM final) of Dexamethasone, 5 mL of 1% Penicillin/Streptomycin (P/S), 72.5 μL (100 nM final) of Insulin.
3. RPMI medium for KC cells: 500 mL of RPMI, 50 mL of 10% Fetal Bovine Serum, 5 mL of 1% Penicillin/Streptomycin (P/S).

2.5 Hepatocyte and KC Immunofluorescence

1. Hepatocyte immunofluorescence: Albumin antibody (R&D Systems) to be used at a concentration of 25 μg/mL (*see* **Note 5**).
2. KC immunofluorescence: F4/80 antibody (AbD Serotec) to be used at a concentration of 1:50.
3. Collagen-coated coverslips for hepatocytes: Coverslip, Rat Tail Collagen Type 1 (*see* **Note 4**).

2.6 Hepatocyte and KC Real-Time PCR Primers

1. Albumin: Forward, TGCTTTTTCCAGGGGTGTGTT; Reverse, TTACTTCCTGCACTAATTTGGCA
2. Clec4f: Forward, GAGGCCGAGCTGAACAGAG; Reverse, TGTGAAGCCACCACAAAAGAG
3. 36B4: Forward, TCCAGGCTTTGGGCATCA; Reverse, CTTT ATCAGCTGCACATCACTCAGA

3 Methods

These procedures should be carried out at room temperature unless otherwise specified.

3.1 Liver Perfusion Reagents

1. 1× HBSS buffer: Freshly prepare 1× HBSS using 10× HBSS and ultrapure water, filter through a 0.22 μm pore size filter.
2. HBSS-EGTA buffer (0.5 mM): Add 500 μL of EGTA 0.5 M to 500 mL of 1× HBSS.
3. HBSS-CaCl$_2$ buffer (1 mM): Add 500 μL of 1 M CaCl$_2$ to 500 mL of 1× HBSS (*see* **Note 1**).

4. Concentrated Collagenase: Add 1.6 g of collagenase from *Clostridium histolyticum* to 50 mL of HBSS-CaCl$_2$ buffer. Filter the solution through a 0.45 μm pore size filter and prepare aliquots of 1 mL that are stored at −20 °C. Dilute 1 mL of concentrated collagenase in 49 mL of HBSS-CaCl$_2$ buffer and heat to 37 °C in water bath before use.

5. Mouse anesthesia: Prepare a solution of 200 μL per mouse of 100 mg/kg of Ketamine and 10 mg/kg of Xylazine in phosphate buffer solution (PBS).

3.2 Percoll Solutions

1. 90% Percoll: For a final volume of 50 mL, add 5 mL of 10× PBS to 45 mL of Percoll.

2. 25% Percoll: For a final volume of 40 mL, add 30 mL of 1× PBS to 10 mL of 90% Percoll.

3. 50% Percoll: For a final volume 30 mL, add 15 mL of 1× PBS to 15 mL of 90% Percoll (*see* **Note 2**).

4. Percoll gradient: In a 50 mL Falcon tube, add 20 mL of 25% Percoll and slowly underlayer it with 14.5 mL of 50% Percoll at the bottom of the tube using a 10 mL plastic pipette (*see* **Note 3**).

3.3 Cultivation Media

1. Adherence medium M199 for hepatocytes: Mix well 475 mL of M199 + Earles Salts, 5 mL of Penicillin/Streptomycin (P/S), 5 mL of BSA (10%), 10 mL of UltroSer™ G (*see* **Note 4**), 5 μL (100 nM final) of Dexamethasone, and 72.5 μL (100 nM final) of Insulin.

2. Basal (maintenance) medium for hepatocytes: Mix well 475 mL of M199 + Earles Salts, 5 μL (100 nM final) of Dexamethasone, 5 mL of 1% Penicillin/Streptomycin (P/S), and 72.5 μL (100 nM final) of Insulin.

3. RPMI medium for KC cells: Mix well 500 mL of RPMI, 50 mL of 10% Fetal Bovine Serum, and 5 mL of 1% Penicillin/ Streptomycin.

3.4 Immuno-fluorescence Antibodies

1. Prepare the albumin antibody (R&D Systems) for the hepatocytes at a concentration of 25 μg/mL.

2. Prepare the F4/80 antibody (AbD Serotec) for the KCs at a concentration of 1:50.

3. Coat the coverslips using Rat Tail Collagen Type 1 (Sigma), for 6 h at RT or overnight at 4 °C. Aspirate the Collagen solution and then dry the coverslips under UV light for 3 h (*see* **Note 6**).

3.5 Mouse Perfusion

1. Warm the HBSS-EGTA buffer, HBSS-CaCl$_2$ buffer, hepatocyte adherence media M199, and RPMI media at 37 °C in a water bath.

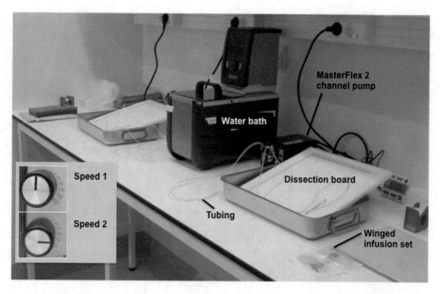

Fig. 1 Setup for mouse liver perfusion

2. Figure 1 illustrates the tools and the equipment to be used for the isolation of KCs and hepatocytes. Prepare all instruments including a dissection board, bulldog clamp (World Precision Instruments), and winged infusion set- 23 G × 0.75 in. needle × 12 in. tubing (BD Safety-Lok™) and 27 G × 1/2 in. (Terumo Corporation). Connect the Tygon® silicone tubing 1/31″ ID 1/8″ OD (Cole-Parmer) to the winged infusion set and pass it through the MasterFlex® 2 channel Pump (Cole-Parmer).

3. Rinse the tubing with PBS, then with HBSS-EGTA for 2 min.

4. Anaesthetize each mouse with Ketamine/Xylazine and wait until it no longer demonstrates a withdrawal reflex (*see* **Note 7**).

5. Place one mouse on its back on a dissection board, wipe the abdomen with ethanol, spread the limbs, and secure them to the dissection board.

6. Make an incision through the skin on the ventral midline and dissect the skin away.

7. Turn on the pump and set it to speed 1 (0.38 mL/min) (*see* Fig. 2a). One end of the tubing should be in the warm HBSS-EGTA and the other should be attached to the winged needle.

8. Open the peritoneal cavity and push the intestines to the right side using a cotton-tip; identify both the inferior vena cava and hepatic portal vein (Fig. 2b).

9. Insert the winged needle in the inferior vena cava. Clamp the vena cava with a bulldog clamp in order to secure the needle

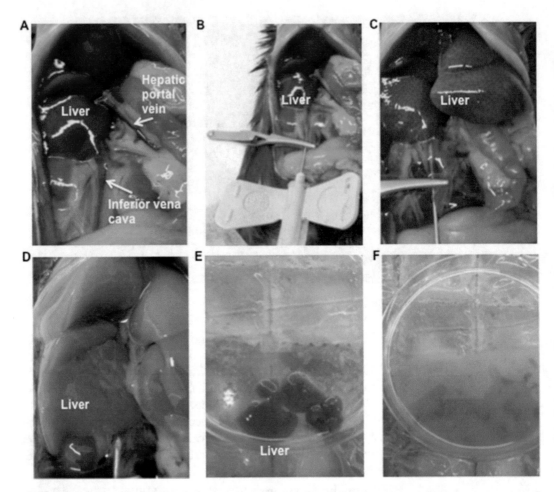

Fig. 2 Step-by-step procedure for hepatic cell isolation from a single mouse liver. (**a**) Inferior vena cava and hepatic portal vein in mouse abdominal cavity. (**b**) Winged needle inserted in the inferior vena cava and secured with a bulldog clamp. (**c**) Initiation of liver discoloration before sectioning the hepatic portal vein. (**d**) Appearance of liver after collagenase digestion. (**e**) Liver dissected after digestion and placed in a Petri dish with cold HBSS-CaCl$_2$ buffer. (**f**) Minced liver

(Fig. 2b). If the perfusion is working, the liver should swell and become discolored (Fig. 2c).

10. Set the pump to speed 2 (1.7 mL/min) (Fig. 2c).

11. Cut the hepatic portal vein. This procedure can also be reversed by inserting the needle into the hepatic portal vein and cutting the inferior vena cava instead (*see* **Note 8**).

12. The liver should now discolor rapidly to a creamy color. If the discoloration is slow or red patches remain, the perfusion did not work and the liver should not be used for the remaining steps.

13. Once the perfusion of the liver has begun, continue to perfuse the liver with approximately 15–20 mL of HBSS-EGTA buffer for approximately 10 min at speed 2.

14. Stop the pump and transfer the tubing to the collagenase buffer.

15. Perfuse the liver until it is sufficiently digested (Fig. 2d). Signs of sufficient digestion are cell release and lobe breakdown when the liver is pressed with a cotton-tip. Digestion of the liver in 10–15 mL of collagenase buffer takes approximately 5–10 min at speed 2 for a lean, healthy mouse. The digestion of the liver for an obese mouse at speed 2 takes approximately 15–30 min and 25–30 mL of collagenase buffer (*see* **Note 9**).

16. After the perfusion is complete, collect the liver and transfer it to a Petri dish containing cold HBSS-CaCl$_2$ buffer. Remove the capsule of Glisson while working on ice (Fig. 2e).

17. Release the hepatic cells gently by cutting the liver lobes and continue until all of the big clumps are gone (Fig. 2f).

18. Filter the cells through a 100 µm cell strainer using 30 mL of cold HBSS-CaCl$_2$ buffer and keep the cells on ice.

19. Centrifuge for 3 min at a speed of 50 × *g* at 4 °C.

20. Collect the supernatant containing the KCs. The hepatocytes are in the pellet (Fig. 3) (*see* **Note 10**).

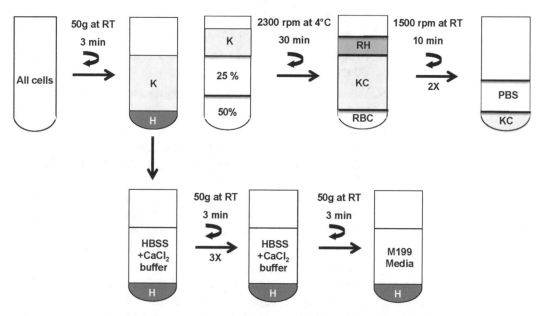

Fig. 3 Step-by-step protocol of hepatic cell fraction separation. *K*, Kupffer cell (KC) enriched fraction, *H*, hepatocyte enriched fraction, *RH*, rest of hepatocytes, *RBC*, red blood cells and cell debris

3.6 Hepatocyte Fraction

1. Gently resuspend the hepatocyte pellet in 30 mL of HBSS-CaCl$_2$, by first resuspending the pellet in a smaller volume of 10 mL of HBSS-CaCl2 with a plastic Pasteur pipette, and then gradually adding the remaining 20 mL of the solution. Centrifuge for 3 min at 50 × g at 4 °C. Aspirate the rinse buffer supernatant.

2. Repeat this step three times.

3. Resuspend the hepatocytes in 30 mL of adherence media M199.

4. Centrifuge for 3 min at 50 × g at 4 °C.

5. Aspirate the supernatant and resuspend the hepatocyte pellet in 10 mL of adherence media M199.

6. Count the cells (*see* **Note 11**). An efficient perfusion should yield approximately 1 × 10^8 cells per liver of an adult mouse.

7. Plate the cells for 3–4 h, and then change the adherence media to the maintenance media (*see* **Note 12**).

3.7 Kupffer Cell Fraction

1. After the first hepatic cell centrifugation, slowly layer the supernatant on the Percoll gradient with a 10 mL pipette and centrifuge the mixture at 1200 × g for 30 min at 4 °C without brakes (*see* **Note 13**).

2. Carefully collect the middle interphase (white cell ring) and transfer it into a new 50 mL tube (*see* Fig. 3).

3. Wash the cells twice by adding PBS up to a final volume of 20 mL and centrifuging at 1500 rpm for 10 min at 4 °C.

4. Resuspend in KC media (*see* **Note 14**) and plate the cells for 30 min.

5. Wash the cells with PBS and add fresh media to remove the non-adherent cells.

6. The purity of the cell fractions can be observed by microscopy using the immunofluorescent antibodies against the macrophage marker F4/80 and the hepatocyte marker albumin (Fig. 4a–b). For quantitative analysis, the expression of the KC marker C-Type Lectin Domain Family 4, Member F (Clec4f) [15] and the hepatocyte marker albumin is measured by real-time PCR (Fig. 4c–d, *see* **Note 14**).

7. Figure 4 shows that all cells in the KC fraction are stained with F4/80 (Fig. 4a), while all cells in the hepatocyte fraction are positive for Albumin (Fig. 4b). Consistent with this observation, *Clec4f* is highly expressed in the KC fraction (Fig. 4c), while A*lb* expression is high in hepatocytes (Fig. 4d).

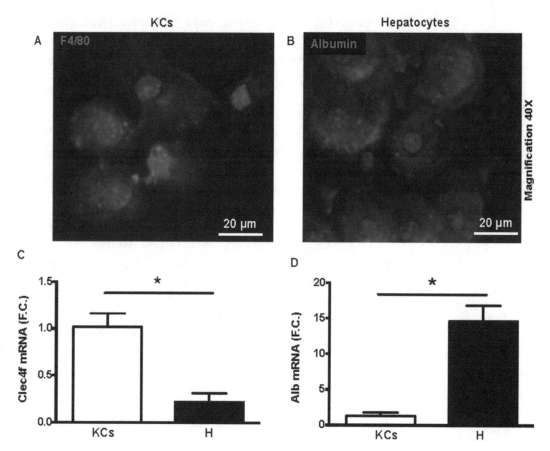

Fig. 4 Purity of isolated Kupffer cell (KCs) and hepatocyte fractions. Cells were isolated from the liver of 14-week-old male C57Bl6/J mice (*n* = 3) on a regular chow diet. (**a**) KCs were stained with an antibody against F4/80 (*red*) and (**b**) hepatocytes were stained with an antibody against albumin (*red*). Nuclei were stained with DAPI (*blue*) (40× magnification). Scale bar, 20 μm. (**c**) Expression of *Clec4f* and (**d**) Albumin (*Alb*) in KC and hepatocyte (H) fractions, values were normalized using the housekeeping gene *36B4*. Results are expressed as the mean fold change (F.C.) value normalized to KC expression ± SEM. Statistical significance was analyzed by Student's *t* test. *$p < 0.05$

All experiments were approved by the Stockholm South Ethical Committee on Animal Research in Huddinge, Sweden.

4   Notes

1. The 500 mL volume of 1xHBSS-CaCl$_2$ is enough for up to three mice. Half of the volume is heated to 37 °C for the perfusion procedure and the other half is kept on ice for the hepatocyte isolation.

2. When preparing the Percoll solution, calculate the required volume accordingly: 1 tube per 1 lean (WT) mouse; and 2 tubes per obese mouse (>35 g). Calculate 2 tubes per

mouse for loading the supernatant suspension from the perfusion procedure.

3. The Percoll gradient needs to be prepared fresh before starting the isolation in order to keep the different layers of the gradient separated.

4. When preparing adherence medium M199 for hepatocytes, UltroSer™ G can be substituted with Fetal Bovine Serum.

5. The Albumin antibody (R&D) has reactivity for both humans and mice.

6. The coverslips can be coated ahead of time, stored at 4 °C, and sealed in Parafilm M®.

7. Lean mice, cca 30 g, take about 10–15 min to get anesthetized; for obese mice it takes approximately 30–40 min.

8. During the perfusion, decrease the rate of the pump to speed 1 to see if the winged infusion set is inserted in the correct position in the vein. If the perfusion does not produce a brighter liver, it is recommended to start over with a new mouse. Nevertheless, the perfusion could continue, but the viability and yield of the cells will be lower compared to a liver that was properly perfused.

9. A longer collagenase digestion of the liver will decrease the cell viability.

10. After centrifugation, work with the cell fractions in sterile conditions in a cell culture hood.

11. Good viability of hepatocytes is above 80%.

12. It is recommended to start experiments 3 h after plating.

13. Use caution while loading the Percoll gradient with supernatant. Use a 10 mL Pasteur pipette with the tip touching the wall of Falcon tube.

14. The KC viability is usually between 90 and 100%. The count varies depending on the mice and the quality of perfusion. Approximately 300,000–500,000 cells can be collected from a liver of C57Bl/6J mice on a regular chow diet, while 1–2 million cells/mouse can be obtained from C57Bl/6J obese mice (>35 g). In addition, flow cytometry analysis shows that 70–80% cells are positive for the F4/80.

Acknowledgments

We are grateful to Ahmed Elewa for critically reading the manuscript. MA and the Karolinska Institutet/AstraZeneca (KI/ZA) Integrated CardioMetabolic Center (ICMC) are supported by funding from AstraZeneca. MA is also supported by the Swedish

Research council. MT is supported by the Novo Nordisk Foundation through the Danish Diabetes Academy and OUH Research grant 15-A845.

References

1. Gao B, Jeong WI, Tian Z (2008) Liver: an organ with predominant innate immunity. Hepatology 47(2):729–736. doi:10.1002/hep.22034

2. Hotamisligil GS (2006) Inflammation and metabolic disorders. Nature 444 (7121):860–867. doi:10.1038/nature05485

3. Racanelli V, Rehermann B (2006) The liver as an immunological organ. Hepatology 43(2 Suppl 1):S54–S62. doi:10.1002/hep.21060

4. Tilg H, Hotamisligil GS (2006) Nonalcoholic fatty liver disease: cytokine-adipokine interplay and regulation of insulin resistance. Gastroenterology 131(3):934–945. doi:10.1053/j.gastro.2006.05.054

5. Smedsrod B, Pertoft H, Gustafson S, Laurent TC (1990) Scavenger functions of the liver endothelial cell. Biochem J 266(2):313–327

6. Brown MS, Goldstein JL (1979) Receptor-mediated endocytosis: insights from the lipoprotein receptor system. Proc Natl Acad Sci U S A 76(7):3330–3337

7. Wang X, Sato R, Brown MS, Hua X, Goldstein JL (1994) SREBP-1, a membrane-bound transcription factor released by sterol-regulated proteolysis. Cell 77(1):53–62

8. Jungermann K, Thurman RG (1992) Hepatocyte heterogeneity in the metabolism of carbohydrates. Enzyme 46(1-3):33–58

9. Knook DL, Sleyster EC (1976) Separation of Kupffer and endothelial cells of the rat liver by centrifugal elutriation. Exp Cell Res 99 (2):444–449

10. Gregory SH, Wing EJ (2002) Neutrophil-Kupffer cell interaction: a critical component of host defenses to systemic bacterial infections. J Leukoc Biol 72(2):239–248

11. Baffy G (2009) Kupffer cells in non-alcoholic fatty liver disease: the emerging view. J Hepatol 51(1):212–223. doi:10.1016/j.jhep.2009.03.008

12. Kolios G, Valatas V, Kouroumalis E (2006) Role of Kupffer cells in the pathogenesis of liver disease. World J Gastroenterol 12 (46):7413–7420

13. Ramadori G, Armbrust T (2001) Cytokines in the liver. Eur J Gastroenterol Hepatol 13 (7):777–784

14. Stienstra R, Saudale F, Duval C, Keshtkar S, Groener JE, van Rooijen N, Staels B, Kersten S, Muller M (2010) Kupffer cells promote hepatic steatosis via interleukin-1beta-dependent suppression of peroxisome proliferator-activated receptor alpha activity. Hepatology 51 (2):511–522. doi:10.1002/hep.23337

15. Tencerova M, Aouadi M, Vangala P, Nicoloro SM, Yawe JC, Cohen JL, Shen Y, Garcia-Menendez L, Pedersen DJ, Gallagher-Dorval K, Perugini RA, Gupta OT, Czech MP (2015) Activated Kupffer cells inhibit insulin sensitivity in obese mice. FASEB J. doi:10.1096/fj.15-270496

16. Tosello-Trampont AC, Landes SG, Nguyen V, Novobrantseva TI, Hahn YS (2012) Kuppfer cells trigger nonalcoholic steatohepatitis development in diet-induced mouse model through tumor necrosis factor-alpha production. J Biol Chem 287(48):40161–40172. doi:10.1074/jbc.M112.417014

17. Petrasek J, Bala S, Csak T, Lippai D, Kodys K, Menashy V, Barrieau M, Min SY, Kurt-Jones EA, Szabo G (2012) IL-1 receptor antagonist ameliorates inflammasome-dependent alcoholic steatohepatitis in mice. J Clin Invest 122(10):3476–3489. doi:10.1172/jci60777

18. Drochmans P, Wanson JC, Mosselmans R (1975) Isolation and subfractionation on ficoll gradients of adult rat hepatocytes. Size, morphology, and biochemical characteristics of cell fractions. J Cell Biol 66(1):1–22

19. Pretlow TG 2nd, Williams EE (1973) Separation of hepatocytes from suspensions of mouse liver cells using programmed gradient sedimentation in gradients of ficoll in tissue sulture medium. Anal Biochem 55(1):114–122

Chapter 17

Alpha-1 Antitrypsin Transcytosis and Secretion

Angelia D. Lockett

Abstract

Protective levels of Alpha-1 antitrypsin (A1AT) are achieved in the lung through the uptake of the pulmonary vasculature of hepatocyte-secreted A1AT. The anti-inflammatory, anti-apoptotic, and anti-protease properties of A1AT are critical toward maintaining the function of pulmonary endothelial and epithelial cells and for the structural integrity of the pulmonary interstitium. To perform these functions A1AT must cross the pulmonary-endothelial barrier. Using transwell inserts, we have demonstrated that the endocytosis of A1AT at the apical surface of endothelial cells, followed by the transcytosis and secretion at the basolateral surface, is a mechanism through which A1AT is transported into the lung epithelium and interstitium.

Key words Western blot, Supernatant, Secretion, Transcytosis, Endocytosis, Transport, Endothelial, Epithelial, Pulmonary

1 Introduction

Alpha-1 antitrypsin (A1AT) is a serine protease inhibitor that is primarily produced in hepatocytes where it undergoes glycosylation and is subsequently secreted into the systemic circulation [1, 2]. A1AT deficiency is a hereditary disease that causes A1AT polymers to accumulate in the liver, thereby leading to a decline in circulating levels as well as deficient lung A1AT levels. Circulating A1AT protects the lung from protease degradation [3], inflammation [4, 5], endothelial cell apoptosis [6, 7], and inflammatory signaling [8]; all of which contribute to the pathogenesis of Chronic Obstructive Pulmonary Disease (COPD). The ability of A1AT to protect endothelial cells is dependent upon the intracellular presence of A1AT, and this process occurs through endothelial cell endocytosis [9, 10] of circulating A1AT that directly contacts the pulmonary vasculature. However, the protection of the lung epithelial cells and alveolar interstitium from proteases requires the passage of A1AT across the pulmonary endothelial barrier.

Florie Borel and Christian Mueller (eds.), *Alpha-1 Antitrypsin Deficiency: Methods and Protocols*, Methods in Molecular Biology, vol. 1639, DOI 10.1007/978-1-4939-7163-3_17, © Springer Science+Business Media LLC 2017

Our studies indicate that in addition to endothelial cell endocytosis of A1AT, there is active transcytosis and directional secretion, which may provide the lung interstitium with protective concentrations of A1AT.

We examined the ability of pulmonary endothelial cells to transcytose and secrete A1AT by culturing confluent monolayers on transwell inserts with a pore size that was sufficient for the passage of proteins and small molecules, but not cells. To insure that we were measuring the active transport of A1AT versus passive leakage, the integrity of the monolayer was tested for leakiness by light microscopy and by treating it with FITC-dextrans [10], (Fig. 1). The coculture of pulmonary endothelial cells on the top of the membrane and pulmonary epithelial cells on the bottom of the membrane, which represents their

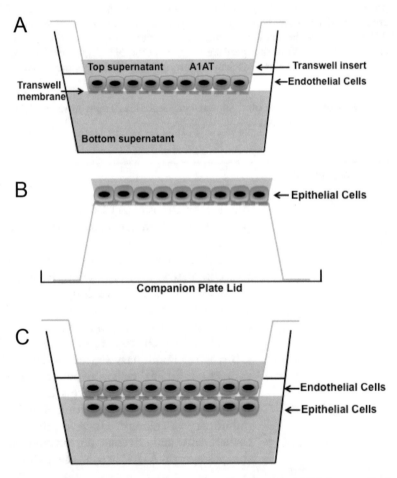

Fig. 1 Integrity of the endothelial monolayer. Fraction (%) of FITC-Dextran (Dex) of 20 kDa (*light gray bars*) or 250 kDa (*dark gray bars*) or of A1AT (100 μg/mL; *black bars*) that crossed confluent endothelial cell monolayers grown on .4 μm transwell inserts. These data were originally published in Plos One [10]

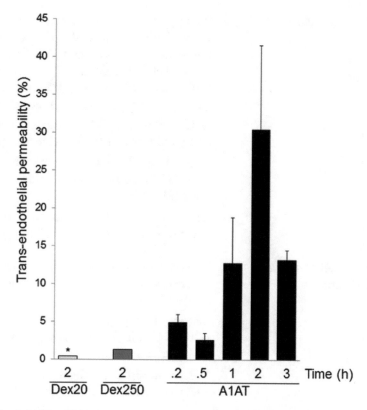

Fig. 2 A1AT trafficking across pulmonary endothelial monolayers and pulmonary endothelial and epithelial cell co-cultures. (**a**) Immunoblots of intracellular and basolaterally secreted A1AT from endothelial cells cultured on transwell inserts and treated with exogenous A1AT (100 μg/mL up to 120 min). (**b**) Immunoblots of A1AT in cell lysates showing the intracellular presence of A1AT in endothelial and epithelial cells, and an immunoblot of secreted A1AT from concentrated bottom supernatants. Bands shown are from the same immunoblot. These data were originally published in Plos One [10]

physiological orientation, demonstrates that A1AT is transcytosed and secreted by the endothelial cells, and then internalized by the epithelial cells (Fig. 2a–b). Through the use of this in vitro coculture system, we have found that endothelial cells take up A1AT and utilize the intracellular protective functions while simultaneously having the capacity to transcytose and secret the protein.

2   Materials

Prepare all buffers and reagents in filtered deionized water (such as from a Millipore filtration system) and store them at room temperature unless otherwise indicated. The primary antibody dilution buffer contains sodium azide. A1AT is prepared from human plasma. Use the solutions with caution and follow the appropriate institutional disposal procedures for all reagents.

2.1 Cell Culture

1. Primary rat lung microvascular endothelial cells (RLMVECs) are available via the University of South Alabama, Center for Lung Biology, Tissue and Cell Culture Core facility (Mobile, AL) and are cultured at 37 °C with 5% CO_2 in 1× DMEM, 1% penicillin/streptomycin, 10% fetal bovine serum.

2. Rat lung epithelial cells, L2, are commercially available from ATCC (Manassas, VA) and are cultured at 37 °C with 5% CO_2, F12 medium, 10% fetal bovine serum.

3. Falcon 6-well transwell inserts with a Polyethylene Terephthalate (PET) membrane containing .4 µM pore size and 6-well companion plates were used.

4. Reconstitute FITC-Dextran in autoclaved filtered deionized water to a concentration of 25 mg/mL.

2.2 A1AT and Protein Concentration

1. Reconstitute purified human A1AT, pooled from human plasma, using autoclaved filtered deionized water to 5 µg/µL concentration.

2. Aliquot 150 µL/tube into eppendorf tubes and store at −20 °C (Avoid freeze-thawing more than two times and thaw on ice prior to use).

3. 3 K Nanosep centrifugal columns to concentrate secreted A1AT come ready to use from Fisher Scientific.

2.3 Lysis Buffer

RIPA buffer stored at 4 °C, 1× Protease Inhibitor Cocktail, 1× PhosStop Cocktail (added immediately before use).

2.4 SDS-PAGE and Transfer

1. Running Buffer: 10× Tris–Glycine SDS Running Buffer: 30 g Tris base, 144 g glycine,10 g SDS in water up to 1 L. 1× running buffer: 100 mL 10× running buffer, 900 mL water.

2. Transfer Buffer: 10× Tris–Glycine Transfer Buffer: 30 g Tris base, 144 g glycine in water up to 1 L. 1× transfer buffer: 100 mL 10× buffer, 700 mL water, 200 mL MeOh.

2.5 Western Blot

1. TBS-T Wash Buffer: Dissolve 1× TBS-T powder containing .05% Tween 20 in 1 L of water and supplement with Tween 20 to a final concentration of .1% (add 500 µL). Alternate is 10× wash buffer stock solution: 24 g Tris base, 87.7 g NaCl in water up to 1 L. Adjust pH to 7.5 with HCl. 1× wash buffer: 100 mL 10× stock, 1 mL Tween-20, 899 mL of water.

2. 5% Blocking Buffer: Dissolve 5 g dry milk powder, 95 mL TBS-T wash buffer.

3. Stock solutions for primary antibody dilution buffer: (a) 0.5 L of .5 M phosphate buffer: Dissolve 8.7 g NaH_2PO_4-H_2O (monobasic), 26.5 g Na_2HPO_4 (dibasic) in water; (b) 0.5 L 5 M NaCl: Dissolve 146.1 g in water; (c) 10 mL 1 M NaN_3:

Dissolve 0.65 g in water; (d) 50 mL 10% Tween 20: 5 mL in 45 mL of water.

4. Primary antibody dilution buffer (0.5 L PDB): 50 mL .5 M phosphate buffer (50 mM final), 15 mL 5 M NaCl (150 mM final), 0.25 mL 10% Tween 20 (.05% final), 0.5 mL 1 M NaN$_3$ (1 mM final), 20 g bovine serum albumin (4% final) in water. Sterile filter using a 0.5 L, .22 μm pore size bottle top filtration unit. Store working solution at 4 °C.

5. Goat anti-human A1AT antibody at 1:1000 dilution in PDB.

6. Anti-goat-HRP-conjugated secondary antibody at 1:5000 dilution in 5% blocking buffer.

3   Methods

3.1 Culturing RLMVECs on Transwell Membranes

1. Heat the supplemented DMEM at 37 °C.

2. Add 2 mL of supplemented DMEM to the bottom chamber.

3. Using sterile forceps, place 6-well transwell inserts into 6-well companion plates, while ensuring that the insert rests between the groves at the top of the well (*see* **Note 1**).

4. Plate 5×10^5 RMLVECs (*see* **Note 2**) per well in 2 mL of supplemented DMEM on the transwell membrane (top chamber) (Fig. 3a).

5. Culture at 37 °C with 5% CO$_2$ for 48 h.

3.2 Verifying Intact (Non-leaky) Monolayer

1. Heat the phenol-free, serum-free (PFSF) DMEM in a 37 °C water bath.

2. Remove the transwell plates from the incubator to the tissue culture hood.

3. Place a pasteur pipet into the groove on the side of the well and aspirate the medium from the bottom chamber.

4. Without disturbing the cell monolayer, aspirate the medium from the top chamber.

5. Rinse the bottom chamber by adding 1 mL of PFSF-DMEM into the groove on the side of the each well and then aspirate it.

6. Rinse the top chamber by carefully adding 1 mL of PFSF-DMEM to the cell monolayer by pipetting it onto the side of the top chamber wall and then aspirate it.

7. Using the groove on the transwell insert, add 1 mL of PFSF-DMEM to the bottom chamber.

8. Carefully add 1 mL of PFSF-DMEM to the cell monolayer by pipetting onto the side of the top chamber wall.

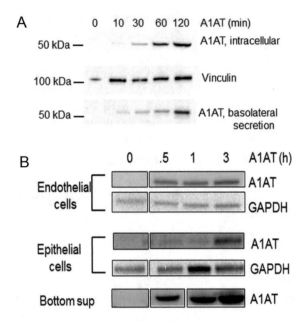

Fig. 3 A schematic of the pulmonary endothelial cell monolayer and pulmonary endothelial and epithelial cell cocultures. (**a**) Depiction of pulmonary endothelial cells cultured in the top chamber of the transwell insert. (**b**) Depiction of a transwell insert flipped onto a companion lid with pulmonary epithelial cells cultured on the bottom of the transwell membrane. (**c**) Depiction of pulmonary endothelial and epithelial cell bilayers. Note that A1AT is added to the top chamber containing endothelial cells

9. Serum starve the cells at 37 °C with 5% CO_2 for 2 h in the tissue culture incubator.

10. Aspirate the media from both chambers as described above.

11. Using the groove on the side of the well, add 1000 μL of PFSF-DMEM to the bottom chamber.

12. Add 960 μL of PFSF-DMEM (1000 μL minus the volume of 20 kDa or 250 kDa FITC-dextran) to the top chamber of the treatment wells.

13. Add 960 μL of PFSF-DMEM (1000 μL minus the volume of the vehicle used for treatments) to the top chamber of the control wells.

14. For the treatment wells, add 1 mg/mL (40 μL) of 20 kDa or 250 kDa FITC-Dextran (*see* **Note 3**).

15. Add 40 μL of the water vehicle to the control wells (*see* **Note 4**).

16. Gently rock the plates side to side to mix.

17. Treat the cells in the tissue culture incubator for 2 h.

18. To measure the permeability of dextran, add 50 μL of the supernatant from the top chamber in duplicate wells of a 96-well plate.

19. Add 50 µL of the supernatant from the bottom chamber into duplicate wells of a 96-well plate.

20. Using a fluorimetric spectrophotometer, measure the fluorescence at an excitation wavelength of 490 nm.

21. Calculate the percent transendothelial permeability by using the following equation:

$$\% \text{ permeability} = \left(\frac{\text{Fluorescence of Bottom supernatant} \times 100}{\text{Fluorescence of Top} + \text{Bottom supernatant}} \right)$$

3.3 Measuring A1AT Transcytosis and Secretion from RLMVECs

1. Culture RLMVECs on transwell membranes as outlined in Subheading 3.1.

2. Heat phenol-containing, serum-free DMEM in a 37 °C water bath.

3.3.1 Serum Starvation and A1AT Treatment

3. Serum starve the cells and prepare them for A1AT treatment by following **steps 1–11** in Subheading 3.2 (use phenol containing DMEM).

4. Add 980 µL of serum-free DMEM (1000 µL minus the volume of A1AT to be used) to the top chamber of treatment wells.

5. Add 980 µL of serum-free DMEM (1000 µL minus the volume of the vehicle used for treatments) to the top chamber of the control wells.

6. For the treatment wells, add 20 µL of (100 µg/mL) A1AT.

7. Add 20 µL of the water vehicle to the control wells.

8. Gently rock the plates side to side to mix.

9. Treat the cells in the tissue culture incubator for 30 min to 2 h.

3.3.2 Harvest Cells and Supernatant

1. Place the transwell plate on ice.

2. Pipet the top supernatant and place it in a 1.5 mL eppendorf tube on ice.

3. Carefully wash the cell monolayer two times with 2 mL of PBS.

4. Place the lid of the plate, inside facing up, on ice.

5. Keeping track of the well order, place the transwell inserts on the lid.

6. Pipet the bottom supernatant and place it in a 1.5 mL eppendorf tube on ice.

7. Add 1 mL of serum-free medium to the cells, scrape gently, and transfer them to an eppendorf tube. Alternatively, scrape in lysis buffer and follow the Lyse Cells procedure that is described below.

8. Centrifuge the top and bottom supernatants and the cells at $500 \times g$ for 5 min at 4 °C.

9. Transfer the top and bottom supernatants to a clean eppendorf tube (if there is a small cell pellet, do not disturb it, discard it) on ice.

10. Aspirate the medium from the scraped cell pellet. The cell pellet can be stored at −80 °C or lysed immediately.

3.3.3 Concentration of Secreted Protein

1. Concentrate 500 μL of the bottom supernatant on a 3 K Nanosep centrifugal column at 16,000 × g at 4 °C for 15–20 min until all the medium is removed (*see* **Note 5**).

2. Wash the column with 20μL of serum-free media and transfer the concentrated protein to a clean eppendorf tube.

3. Store the concentrated protein at −80 °C.

3.3.4 Lysis of Cells

1. Add 200 μL of RIPA buffer (containing 1× protease and 1× phosphatase inhibitors) to the cell pellet.

2. Pipet the cells in order to resuspend them. Vortex for 10 s.

3. Incubate the cells on ice for 30 min.

4. Sonicate for 5 s on ice at power level 5.

5. Centrifuge at 4 °C for 10 min at 16,000 × g.

6. Transfer the whole cell lysate to a clean eppendorf tube.

7. Store whole cell lysates at −80 °C.

3.3.5 Prepare Samples for SDS-PAGE

1. Thaw the cell lysates on ice.

2. Determine the protein concentration by performing a protein assay (i.e., BCA) on whole cell lysates.

3. Add 30 μg of the whole cell lysate to a clean eppendorf tube and add the appropriate volume of laemmli buffer (containing 10% 2-mercaptoethanol) to obtain a final buffer concentration of 1×.

4. Add the appropriate volume of laemmli buffer to obtain a final buffer concentration of 1×, to the concentrated bottom supernatant from Subheading 3.3.3, **step 2**.

5. Boil the bottom supernatants and whole cell lysates for 5 min.

6. The samples can be stored at −20 °C or loaded immediately onto a gel (centrifuge samples before loading the gel).

3.3.6 SDS-PAGE/ Transfer

1. Load the whole cell lysates and bottom supernatants unto a 4–20% Tris–HCl Bio-rad Criterion gel (can also use a 10% Tris–HCl gel).

2. Run the gel at 170 V for 1 h at room temperature.

3. Transfer at 100 V for 30 min at 4 °C with an ice block.

3.3.7   Western Blot

1. Block the membrane on a rocker for 1 h at room temperature in 5% milk solution (*see* Subheading 2.5).

2. Pour off the milk and perform one quick rinse with TBS-T.

3. Wash membrane three times, for 10 min each, with TBS-T.

4. Dilute the goat anti-human A1AT antibody 1:1000 in PDB.

5. Incubate the membrane on a rocker at room temperature for 1 h.

6. Pour off the antibody (the antibody can be stored at 4 °C and reused) and wash the membrane as outlined in **steps 2** and **3**.

7. Dilute anti-goat-HRP-conjugated secondary antibody 1:5000 in 5% milk solution.

8. Incubate the membrane on a rocker at room temperature for 1 h.

9. Pour off milk and repeat wash steps as outlined in **steps 2** and **3**.

10. Add the chemiluminescence reagents to the membrane and develop (*see* **Note 6**).

3.4   Coculturing RLMVECs and Lung Epithelial Cells on Transwell Membranes

1. Heat supplemented F12 medium (10% fetal bovine serum) at 37 °C.

2. Place the lid of a 6-well companion plate on the culture hood surface with the inside of the lid facing upward.

3.4.1   Culturing Lung Epithelial Cells

3. Using sterile forceps, place the 6-well transwell inserts upside down (with the bottom of the membrane facing upward) onto the companion plate lid (Fig. 3b).

4. Pipet 1 mL of supplemented F12 medium containing 2.5×10^5 cells onto the middle of the transwell membrane.

5. Flip the bottom of the companion plate upside down (with the wells facing downward), and carefully place it onto the transwell insert membrane. The medium should spread to the edge of the membrane without spilling over the side of the insert.

6. Place the upside down transwell plate and the inserts in a tissue culture incubator at 37 °C for 3 h in order to allow the cells to attach to the membrane.

3.4.2   Coculture of RLMVECs

1. Heat supplemented DMEM in a 37 °C water bath.

2. Remove the transwell plate from the incubator and while maintaining the upside down position, place the plate in a culture hood.

3. Remove the bottom of the companion plate and place it upright in the culture hood.

4. Pipet 2 mL of supplemented F12 medium into each companion plate well.

5. Aspirate the medium from the transwell membrane.

6. Using sterile forceps, place the transwell inserts into the companion plates, while ensuring that the insert rests between the groves at the top of the well.

7. Plate 5×10^5 RMLVECs per well in 2 mL of supplemented DMEM on the transwell membrane (top chamber) (Fig. 3c).

8. Culture at 37 °C for 48 h.

3.5 Measuring A1AT Transcytosis and Secretion in Cocultures

3.5.1 Serum Starvation

1. Heat phenol-containing, serum-free DMEM and serum-free F12 media in a 37 °C water bath.

2. Place the cocultured transwell plate in the tissue culture hood.

3. Place a pasteur pipet into the groove on the side of the well and aspirate the medium from the bottom chamber.

4. Without perturbing the cell monolayer, aspirate the medium from the top chamber.

5. Using the groove on the transwell insert, add 1 mL of serum-free F12 medium to the bottom chamber.

6. Carefully add 1 mL of serum-free DMEM to the cell monolayer by pipetting it onto the side of the top chamber wall.

7. Serum starve the cells at 37 °C for 2 h in the tissue culture incubator.

3.5.2 A1AT Treatment

1. Aspirate the media from both chambers as described in **steps 3–4** of Subheading 3.5.1.

2. Using the groove on the transwell insert, add 1 mL of serum-free F12 medium to the bottom chamber.

3. Add 980 µL of serum-free DMEM (1000 µL minus the volume of A1AT to be used) to the treatment wells in the top chamber.

4. Add 980 µL of serum-free DMEM (1000 µL minus the volume of vehicle used for treatments) to the control wells in the top chamber.

5. For the treatment wells, add 20 µL of A1AT (final 100 µg/mL).

6. Add 20 µL of the water vehicle to control wells.

7. Gently rock the plates side to side to mix.

8. Treat the cells in tissue culture incubator for 30 min to 2 h.

3.5.3 Harvest Cells and Supernatant

1. Follow Subheading 3.3.2, **steps 1–10**, in order to harvest the top and bottom supernatants and the endothelial cells.

2. To harvest the epithelial cells, flip the transwell insert upside down on the plate lid on ice.

3. Add 200 µL of RIPA buffer to the epithelial cells.

4. Gently scrape the cells and transfer them to a 1.5 mL Eppendorf tube on ice.

5. Follow Subheadings 3.3.3–3.3.7 to concentrate the supernatant, isolate whole cell lysates, and perform a Western blot for A1AT.

4 Notes

1. Each side of the well has a groove for the hinge of the transwell membrane to fit into, otherwise the membrane will not be submerged in the bottom medium.

2. This protocol has been optimized for RLMVECs, for other cell types, follow the methods that are described in Subheading 3.2.

3. A1AT and Dextran uptake can be compared by either labeling A1AT with Alexa Fluor 488 (Invitrogen: follow Subheading 3.2, **steps 18–21** to measure the permeability) or by adding unlabeled A1AT and following Subheading 3.3 to measure A1AT secretion and transcytosis. A1AT will also need to be measured by Western blot in the top supernatant (A1AT is highly concentrated in the top supernatant, 10 μL is sufficient for a Western Blot). The %permeability is calculated as follows: %permeability = density of A1AT in bottom supernatant/density of all A1AT (top supernatant + intracellular + bottom supernatant) × 100. Using this method we have previously observed 30% permeability of A1AT versus 1.5% of FITC Dextran Fig. 1.

4. As an optional control, prepare one well containing medium in the top and bottom chambers, but do not add cells to the transwell membrane. Add FITC-Dextran and measure the permeability as described. This controls for pore size and the ability of FITC-Dextran to cross the membrane.

5. To normalize for equal loading among samples containing secreted A1AT, it is ideal to remove all of the medium from the column and resuspend the samples in equal volumes. However, sometimes, the column will not remove all the medium. In such a case, either load the total volume of each sample for the Western blot or use a pipet to determine the volume of each sample and use an equal percentage of each.

6. If using film, ECL is sufficient to detect A1AT. If using imaging equipment such as the Bio-rad Chemidoc, use a stronger reagent such as ECL Plus.

References

1. Vaughan L, Lorier MA, Carrell RW (1982) Alpha 1-antitrypsin microheterogeneity. Isolation and physiological significance of isoforms. Biochim Biophys Acta 701(3):339–345

2. Mills K, Mills PB, Clayton PT, Johnson AW, Whitehouse DB, Winchester BG (2001) Identification of alpha(1)-antitrypsin variants in plasma with the use of proteomic technology. Clin Chem 47(11):2012–2022

3. Janoff A (1985) Elastases and emphysema. Current assessment of the protease-antiprotease hypothesis. Am Rev Respir Dis 132(2):417–433

4. Churg A, Wang X, Wang RD, Meixner SC, Pryzdial EL, Wright JL (2007) Alpha1-antitrypsin suppresses TNF-alpha and MMP-12 production by cigarette smoke-stimulated macrophages. Am J Respir Cell Mol Biol 37 (2):144–151. doi:10.1165/rcmb.2006-0345OC. 2006-0345OC [pii]

5. Churg A, Wang RD, Xie C, Wright JL (2003) Alpha-1-antitrypsin ameliorates cigarette smoke-induced emphysema in the mouse. Am J Respir Crit Care Med 168(2):199–207

6. Petrache I, Fijalkowska I, Zhen L, Medler TR, Brown E, Cruz P, Choe KH, Taraseviciene-Stewart L, Scerbavicius R, Shapiro L, Zhang B, Song S, Hicklin D, Voelkel NF, Flotte T, Tuder RM (2006) A novel anti-apoptotic role for alpha-1 antitrypsin in the prevention of pulmonary emphysema. Am J Respir Crit Care Med 173(11):1222–1228

7. Petrache I, Fijalkowska I, Medler TR, Skirball J, Cruz P, Zhen L, Petrache HI, Flotte TR, Tuder RM (2006) {alpha}-1 antitrypsin inhibits caspase-3 activity, preventing lung endothelial cell apoptosis. Am J Pathol 169 (4):1155–1166

8. Lockett AD, Kimani S, Ddungu G, Wrenger S, Tuder RM, Janciauskiene SM, Petrache I (2013) Alpha(1)-antitrypsin modulates lung endothelial cell inflammatory responses to TNF-alpha. Am J Respir Cell Mol Biol 49 (1):143–150. doi:10.1165/rcmb.2012-0515OC

9. Sohrab S, Petrusca DN, Lockett AD, Schweitzer KS, Rush NI, Gu Y, Kamocki K, Garrison J, Petrache I (2009) Mechanism of {alpha}-1 antitrypsin endocytosis by lung endothelium. FASEB J. doi:10.1096/fj.09-129304. fj.09-129304 [pii]

10. Lockett AD, Brown MB, Santos-Falcon N, Rush NI, Oueini H, Oberle AJ, Bolanis E, Fragoso MA, Petrusca DN, Serban KA, Schweitzer KS, Presson RG, Jr., Campos M, Petrache I (2014) Active trafficking of alpha 1 antitrypsin across the lung endothelium. PLoS One 9 (4):e93979. doi:10.1371/journal.pone.0093979

Chapter 18

Measuring the Effect of Histone Deacetylase Inhibitors (HDACi) on the Secretion and Activity of Alpha-1 Antitrypsin

Chao Wang, Marion Bouchecareilh, and William E. Balch

Abstract

Alpha-1 antitrypsin deficiency (AATD) is a protein conformational disease with the most common cause being the Z-variant mutation in alpha-1 antitrypsin (Z-AAT). The misfolded conformation triggered by the Z-variant disrupts cellular proteostasis (protein folding) systems and fails to meet the endoplasmic reticulum (ER) export metrics, leading to decreased circulating AAT and deficient antiprotease activity in the plasma and lung. Here, we describe the methods for measuring the secretion and neutrophil elastase (NE) inhibition activity of AAT/Z-AAT, as well as the response to histone deacetylase inhibitor (HDACi), a major proteostasis modifier that impacts the secretion and function of AATD from the liver to plasma. These methods provide a platform for further therapeutic development of proteostasis regulators for AATD.

Key words Alpha-1 antitrypsin deficiency, Cellular proteostasis, Histone deacetylase inhibitor, Proteostasis regulators

1 Introduction

Alpha-1 antitrypsin deficiency (AATD) is a hereditary disorder caused by mutations in the Alpha-1 antitrypsin (AAT) gene [1]. AAT is the most abundant serine protease inhibitor (SERPIN) in the plasma that is synthesized in hepatocytes, and it plays a critical role in preventing the degradation of lung tissue by neutrophil elastase (NE). AAT is considered a metastable protein that is managed by a proteostasis system which encompasses protein synthesis, folding, degradation, and membrane trafficking systems [2–5]. This evolutionarily ancient system optimizes biological processes which facilitate the synthesis, processing, and maintenance of protein folds that elicit specific functions in response to different cellular environments [6]. However, AAT variants arising during the normal evolution of the genome may exceed proteostasis capacity and trigger disease phenotypes.

Florie Borel and Christian Mueller (eds.), *Alpha-1 Antitrypsin Deficiency: Methods and Protocols*, Methods in Molecular Biology, vol. 1639, DOI 10.1007/978-1-4939-7163-3_18, © Springer Science+Business Media LLC 2017

The most common AAT variant that causes AATD is the Z-variant (E342K). This variant has a slow folding rate and an increased tendency to form polymers through a "loop-sheet" mechanism in the endoplasmic reticulum (ER) [7]. Z-AAT polymers result in proteotoxic stress in the ER, which triggers liver diseases such as neonatal hepatitis, cirrhosis, and hepatocellular carcinoma [8]. The misfolded Z-AAT conformation also decreases its trafficking efficacy along the secretory pathway, and therefore leads to a substantially reduced level of circulating AAT in the plasma that is required for antiprotease activity in the lung. Loss of functional AAT in the lung leads to emphysema and/or chronic obstructive pulmonary disease (COPD) [9]. Thus, AATD is considered to be a "protein conformational disease" [10, 11]. Understanding and managing the balance between evolutionarily diverse protein fold trajectories and the proteostasis system's ability to regulate the folding, trafficking, and function of proteins represents a largely untested mechanism for disease intervention in terms of the development of proteostasis targeted therapeutics [12].

The acetylation and deacetylation status of protein lysine residues, which is mediated by histone acetyltransferases (HATs) and histone deacetylases (HDACs), has a major impact on protein folding properties and the proteostasis system [13]. It is well known that this modification of histone proteins controls the nucleosome structure and regulates gene expressions through epigenetic processes. Acetylation and deacetylation also impacts almost all proteostasis components and pathways, including: (a) the core chaperones that are responsible for assisting protein folding (e.g., Hsp90 [14], Hsp70 [15]; Hsp40 [16], Bip [17]; (b) the transcription factor HSF1, which is critical for stress response [18]; (c) Lys ubiquitination components (by direct competition), which are central in degradation processes [19]; and (d) cytoskeletal components that are responsible for membrane trafficking processes [20]. The key role of HATs and HDACs in proteostasis is further supported by the beneficial effects of HDAC inhibitors (HDACi) in many protein misfolding diseases, such as cystic fibrosis [21], Gaucher's disease [22], muscle atrophy [23], Niemann–Pick C [24], and neurodegenerative diseases [25].

We previously showed that HDACi suberoylanilide hydroxamic acid (SAHA) improved the secretion of functional of Z-AAT from less than 20% to approximately 50% in a Wild-Type (WT-AAT) cell line [26]. Here, we describe the detailed methods for measuring the HDACi effect on both the trafficking efficacy and NE inhibition activity of AAT and Z-AAT. AAT has three N-linked glycosylation sites. The processing states of oligosaccharides can demonstrate the trafficking efficacy of immature, core-glycosylated isoforms that reflect ER fractions as well as mature, complex-glycosylated isoforms that reflects Golgi fractions [27]. The mature isoform that is secreted from cells shows slower migration during sodium dodecyl sulfate polyacrylamide gel electrophoresis (SDS-PAGE) than the immature isoform. In order to inhibit NE, AAT forms a covalent and irreversible complex with NE, which can be measured by SDS-PAGE as a readout for AAT function [28]. Therefore, this type of assay can serve as a platform to test proteostasis compounds that have therapeutic potential for AATD such as HDACi, which is described here.

2 Materials

2.1 Cellular Sample Materials

1. Cell lines: Human epithelial colorectal carcinoma cell line (HCT116) (see **Notes 1** and **2**).
2. Culture medium: Dulbecco's Modified Eagle's Medium (DMEM), fetal bovine serum (FBS).
3. Suberoylanilide hydroxamic acid (SAHA) solution (Cayman Chemical, Ann Arbor, MI, USA): SAHA, 50 mM dimethyl sulfoxide (DMSO).
4. Cell culture equipment: Biosafety hood, 70% ethanol, incubator maintained at 37 °C and 5% CO_2, water bath, microscope, 12-well cell culture dish, pipettes, pipettor.
5. 1× PBS buffer: 137 mM NaCl, 2.7 mM KCl, 1.8 mM KH_2PO_4, 15.2 mM $Na_2HPO_4/7H_2O$.
6. Cell lysis buffer: 50 mM Tris–HCl (pH 7.4), 150 mM NaCl, 1% (v/v) Triton X-100 containing protease inhibitor cocktail at 2 mg/mL.
7. Centrifuge capable of holding 1.5 mL tubes and reaching centrifugal force of 20,000 × g at 4 °C.

2.2 Immunoblotting AAT/Z-AAT

1. Bradford protein assay kit.
2. 8% and 10% SDS-PAGE gels.
3. 6× SDS-PAGE loading buffer: 30% glycerol, 120 mM Tris pH 7.0, 6% SDS, 0.6% Bromophenol Blue with 6% β-mercaptoethanol.
4. 1× running buffer: 25 mM Tris, 192 mM glycine, 0.1% SDS, SeeBlue® prestained protein standard (Life Technologies, Grand Island, NY, USA).
5. SDS-PAGE materials: heating block, electrode/running tank.
6. Protein transfer materials: nitrocellulose membrane, wet-electroblotting systems (transfer tank, cassette, blot paper, glass tube, fiber pad, ice packs, glass dish, stir bar.
7. 1× transfer buffer: 25 mM Tris, 192 mM glycine.
8. Antibodies: AAT antibody (Immunology Consultants Laboratory, Inc. Portland, OR, USA), Hsp90 antibody (Enzo Life

Sciences, Plymouth Meeting, PA, USA.), peroxidase conjugated secondary antibody (mouse anti-goat IgG for AAT primary antibody, goat anti-rabbit IgG for Hsp90 antibody).

9. 1× TBST buffer: 25 mM Tris, 150 mM NaCl, 2 mM KCl, pH 7.4, 0.1% Tween 20.

10. Nonfat dry milk.

11. ECL Solution-1: 2.5 mM luminol, 0.45 mM p-coumaric acid, 0.1 M Tris pH 8.8.

12. ECL Solution-2: 0.02% hydrogen peroxide, 0.1 M Tris pH 8.8.

13. Film, developing cassette, dark room, developer.

2.3 Detecting the Complex of AAT/Z-AAT and NE

1. Human neutrophil elastase (Innovative research, MI, USA).

2. Phenylmethanesulfonylfluoride (PMSF) solution: PMSF, isopropanol.

3 Methods

3.1 Cell Culture Solution and AAT/Z-AAT Detection Solution

1. In order to prepare a stock solution for cell culture, dissolve SAHA in 50 mM DMSO of and store in aliquots at −20 °C.

2. In order to prepare a stock solution for AAT/Z-AAT detection, dissolve PMSF in isopropanol to a concentration of 20 mM (20×) and store it at −20 °C.

3.2 Cell Culture, Compound Treatment, and Sample Collection

1. Plate 10×10^4 WT-AAT and Z-AAT HCT116 cells in a 12-well tissue culture dish with DMEM containing 10% FBS and let them grow to confluent state (usually 2–3 days) (*see* **Note 3**).

2. When the cells are confluent, renew the culture medium (DMEM containing 10% FBS).

3. Add SAHA to a final concentration of 5 μM to the compound treated wells and an equal volume of DMSO for control wells (*see* **Note 4**).

4. Incubate the cells at 37 °C for 24 h.

5. At the end of the treatment, remove and discard the culture medium. Wash the cells twice with 1× PBS.

6. Add 350 μM of serum-free medium (only DMEM) to collect the secreted AAT/Z-AAT.

7. Incubate the cells at 37 °C for 2 h.

8. After 2 h, harvest the culture medium that contains the secreted AAT/Z-AAT (*see* **Note 5**).

9. Samples can be aliquoted and stored at −80 °C.

10. Put the culture dish on ice and wash the dish twice with cold 1× PBS.

11. Add 50 µL lysis buffer to each well.

12. Lyse the cells for 30 min on ice and rock the dish every 10 min.

13. After 30 min, scrape the wells and transfer the lysate to 1.5 mL centrifuge tubes.

14. Spin the lysate at 20,000 × *g* for 20 min at 4 °C.

15. Collect the supernatant of the lysate and transfer it to a new set of tubes (*see* **Note 6**).

16. Samples can be stored at −80 °C.

3.3 Measuring the Secretion of AAT/Z-AAT in Response to SAHA

1. Determine the protein concentration of the cell lysate supernatant by using the Bradford protein assay kit.

2. Prepare loading samples by adding SDS-PAGE loading buffer and heating the sample at 95 °C for 5 min (*see* **Note 7**).

3. Load 15 µg lysate supernatant samples to a 10% SDS-PAGE gel for probing the immature and mature AAT/Z-AAT.

4. Load 20 µL collected culture medium samples to another 10% SDS-PAGE gel for probing the secreted AAT/Z-AAT.

5. Load the prestained protein standard ladder on both gels.

6. Run the gel at 30 mA/gel and 200 V until the 50 kDa marker band nears the bottom (AAT is 52 kDa).

7. Stop the gel and perform a protein transfer to a nitrocellulose membrane at 500 mA of current and 100 V for 1.5 h.

8. Hsp90 is the loading control. The immature, mature, and secreted bands of AAT/Z-AAT (52 kDa of molecular weight) are found between the markers of 50 kDa and 64 kDa (See-Blue® prestained protein standard). When considering other proteins as loading controls, *see* **Notes 8** and **9**.

9. Cut the blot above 64 kDa in order to separate the intracellular AAT/Z-AAT from Hsp90.

10. Block the blots in 5% milk (diluted in 1× TBST) at room temperature for 1 h.

11. Dilute the AAT antibody and Hsp90 antibody in 1% milk with 1:2000 and 1:25,000 ratios, respectively.

12. Incubate the blots in the primary antibody at 4 °C on the rocker platform overnight.

13. On the next day, discard the primary antibody and wash the blots with 1× TBST buffer for 10 min three times.

14. Then incubate the blots in the peroxidase-conjugated secondary antibody (1:10,000 diluted in 1% milk) for 1 h at room temperature.

190 Chao Wang et al.

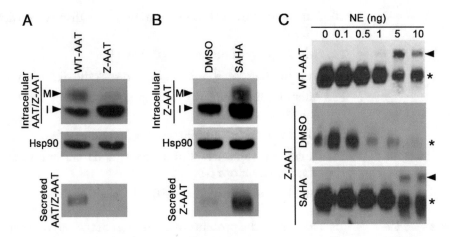

Fig. 1 Western blot shows the secretion and NE inhibition activity of Z-AAT with or without SAHA treatment. (**a**) Secretion of WT-AAT or Z-AAT. *M* mature fraction, *I* immature fraction. (**b**) Secretion of Z-AAT under DMSO or SAHA condition. (**c**) NE inhibition activity of WT-AAT and Z-AAT with or without SAHA treatment. The band indicated by *arrow* is the AAT-NE complex. The band labeled by *asterisk* is the unbounded AAT. This figure is derived from reference [26]

15. Discard the secondary antibody and wash the blots with 1× TBST buffer for 10 min three times.

16. After the wash, dab the blots on filter paper; place the blots on plastic wrap, add 2 mL of the ECL mixture (mix equal volumes of solution 1 and 2) to each blot, and incubate for 1 min.

17. Dab the excess liquid off and put blots in the developing cassette with the transparent sheet protector.

18. Perform the film exposure and developing in the dark room.

As shown in Fig.1a, the mature fraction (M) of WT-AAT migrates slower than the immature fraction (I). Compared to WT-AAT, the mature and secreted fraction of Z-AAT is significantly smaller, while the immature fraction is larger, indicating Z-AAT is largely retained in the ER. SAHA significantly increases both the mature and secreted fraction (Fig. 1b), indicating the correction effect of HDACi on the deficient secretion of Z-AAT.

3.4 Measuring the Antiprotease of AAT/Z-AAT in Response to SAHA

1. Prepare NE solution in 1× PBS buffer at 0.1, 0.5, 1, 2, 5, and 10 ng/μL.

2. Add 1 μL of each of the NE solutions to 20 μL of the culture medium that were collected in each experiment that contains secreted AAT or Z-AAT.

3. Tap the tubes several times to mix the solution, then briefly spin down the mixture.

4. Incubate the tubes at 37 °C for 30 min in order to perform the binding reaction.

5. After the reaction is complete, add SDS-PAGE loading buffer and heat the samples at 95 °C for 5 min (*see* **Note 10**).

6. Run the samples by using 8% SDS-PAGE and follow the same western blot procedure described above. Use the AAT antibody to detect the AAT-NE complex.

As shown in Fig.1c, the covalent complex between WT-AAT and NE is observed when 5 ng of NE was added in the solution (the band is indicated by the arrow). No complex is observed for Z-AAT. In contrast, after SAHA treatment, we observed the complex between Z-AAT and NE, which indicates that SAHA induced an increase in the antiprotease activity of secreted Z-AAT.

4 Notes

1. Although AAT is mainly synthesized and secreted from hepatocytes, the large amount of endogenous WT-AAT limits the use of the liver cell line to study the biology of the exogenously expressed Z-variant AAT. Therefore, we used a human epithelial colorectal carcinoma cell line (HCT116) with undetectable endogenous AAT to generate a stable cell line that expresses the exogenous WT-AAT or Z-AAT with FLAG and HA tags at the N-terminus [26].

2. All the materials listed need to be prepared in an aseptic environment or sterilized in order to avoid contamination.

3. For HCT116 cell maintenance, the medium (DMEM containing 10% FBS) needs to be renewed every 2 to 3 days, and a subcultivation ratio of 1:3 to 1:8 is recommended. The culture environment is 37 °C with 95% air and 5% carbon dioxide.

4. A dose-dependent increase effect of SAHA was observed in our previous study [26]. A dose of 0.5 μM of SAHA starts to show a corrective effect, and a dose of 5 μM of SAHA produces the maximal corrective effect. A dose higher than 5 μM is toxic to the cell.

5. After collecting the culture medium that contains secreted AAT/Z-AAT, it is better to add fresh medium back to the cells in order to prevent drying. This also provides more time to prepare the next step.

6. The pellet of the cell lysate can be kept for the analysis of aggregates of AAT/Z-AAT [29].

7. The loading volume of the collected culture medium can be normalized according to the protein concentration of the cell lysate.

8. Actin (43 kDa for either alpha or beta subunit) and tubulin (55 kDa for either alpha or beta subunit) are not recommended

for use as loading controls because they have molecular weights that are similar to AAT (52 kDa) and cannot be separated from AAT in SDS-PAGE.

9. GAPDH (36 kDa) can be used as a loading control for intracellular AAT. If GAPDH is used as a loading control, make sure that (a) GAPDH does not run out of the gel, and (b) cut the membrane below the 50 kDa marker band carefully to separate AAT and GAPDH. Ponceau staining of the membrane can provide a general view of how the samples run on the SDS-PAGE, which helps facilitate proper cutting.

10. It is recommended to add 1 mM of PMSF to stop the NE binding reaction before adding the SDS-PAGE loading buffer. After forming the covalent complex with NE, AAT undergoes large conformational changes and becomes disordered in some regions, which makes the complex very susceptible to being digested by protease, including NE. By adding PMSF immediately after the reaction, it is possible to detect the very weak Z-AAT and NE complex (data not shown).

Acknowledgments

This work was supported by grants from National Institute of Health HL095524 for WEB. CW is supported by a postdoctoral research fellowship from Alpha-1 Foundation.

Reference

1. Ghouse R, Chu A, Wang Y, Perlmutter DH (2014) Mysteries of alpha1-antitrypsin deficiency: emerging therapeutic strategies for a challenging disease. Dis Model Mech 7:411–419

2. Balch WE, Morimoto RI, Dillin A, Kelly JW (2008) Adapting proteostasis for disease intervention. Science 319:916–919

3. Bouchecareilh M, Conkright JJ, Balch WE (2010) Proteostasis strategies for restoring alpha1-antitrypsin deficiency. Proc Am Thorac Soc 7:415–422

4. Bouchecareilh M, Balch WE (2011) Proteostasis: a new therapeutic paradigm for pulmonary disease. Proc Am Thorac Soc 8:189–195

5. Bouchecareilh M, Balch WE (2012) Proteostasis, an emerging therapeutic paradigm for managing inflammatory airway stress disease. Curr Mol Med 12:815–826

6. Powers ET, Balch WE (2013) Diversity in the origins of proteostasis networks–a driver for protein function in evolution. Nat Rev Mol Cell Biol 14:237–248

7. Gooptu B, Dickens JA, Lomas DA (2014) The molecular and cellular pathology of alpha(1)-antitrypsin deficiency. Trends Mol Med 20:116–127

8. Teckman JH (2013) Liver disease in alpha-1 antitrypsin deficiency: current understanding and future therapy. COPD 10(Suppl 1):35–43

9. Brebner JA, Stockley RA (2013) Recent advances in alpha-1-antitrypsin deficiency-related lung disease. Expert Rev Respir Med 7:213–229; quiz 230

10. Kopito RR, Ron D (2000) Conformational disease. Nat Cell Biol 2:E207–E209

11. Nyon MP, Gooptu B (2014) Therapeutic targeting of misfolding and conformational change in alpha1-antitrypsin deficiency. Future Med Chem 6:1047–1065

12. Wang C, Balch WE (2015) Managing the adaptive proteostatic landscape: restoring resilience in alpha-1 antitrypsin deficiency. In: Sandhaus RA, Wanner A (eds) Alpha-1 antitrypsin: role in health and disease. Springer, Cham, Switzerland. (in press)

13. Hutt DM, Balch WE (2013) Expanding proteostasis by membrane trafficking networks. Cold Spring Harb Perspect Med 3:1–21

14. Kovacs JJ, Murphy PJ, Gaillard S, Zhao X, Wu JT, Nicchitta CV, Yoshida M, Toft DO, Pratt WB, Yao TP (2005) HDAC6 regulates Hsp90 acetylation and chaperone-dependent activation of glucocorticoid receptor. Mol Cell 18:601–607

15. Marinova Z, Ren M, Wendland JR, Leng Y, Liang MH, Yasuda S, Leeds P, Chuang DM (2009) Valproic acid induces functional heat-shock protein 70 via Class I histone deacetylase inhibition in cortical neurons: a potential role of Sp1 acetylation. J Neurochem 111:976–987

16. Hageman J, Rujano MA, van Waarde MA, Kakkar V, Dirks RP, Govorukhina N, Oosterveld-Hut HM, Lubsen NH, Kampinga HH (2010) A DNAJB chaperone subfamily with HDAC-dependent activities suppresses toxic protein aggregation. Mol Cell 37:355–369

17. Rao R, Nalluri S, Kolhe R, Yang Y, Fiskus W, Chen J, Ha K, Buckley KM, Balusu R, Coothankandaswamy V, Joshi A, Atadja P, Bhalla KN (2010) Treatment with panobinostat induces glucose-regulated protein 78 acetylation and endoplasmic reticulum stress in breast cancer cells. Mol Cancer Ther 9:942–952

18. Westerheide SD, Anckar J, Stevens SM Jr, Sistonen L, Morimoto RI (2009) Stress-inducible regulation of heat shock factor 1 by the deacetylase SIRT1. Science 323:1063–1066

19. Wagner SA, Beli P, Weinert BT, Nielsen ML, Cox J, Mann M, Choudhary C (2011) A proteome-wide, quantitative survey of in vivo ubiquitylation sites reveals widespread regulatory roles. Mol Cell Proteomics 10:M111 013284

20. Janke C, Bulinski JC (2011) Post-translational regulation of the microtubule cytoskeleton: mechanisms and functions. Nat Rev Mol Cell Biol 12:773–786

21. Hutt DM, Herman D, Rodrigues AP, Noel S, Pilewski JM, Matteson J, Hoch B, Kellner W, Kelly JW, Schmidt A, Thomas PJ, Matsumura Y, Skach WR, Gentzsch M, Riordan JR, Sorscher EJ, Okiyoneda T, Yates JR 3rd, Lukacs GL, Frizzell RA, Manning G, Gottesfeld JM, Balch WE (2010) Reduced histone deacetylase 7 activity restores function to misfolded CFTR in cystic fibrosis. Nat Chem Biol 6:25–33

22. Lu J, Yang C, Chen M, Ye DY, Lonser RR, Brady RO, Zhuang Z (2011) Histone deacetylase inhibitors prevent the degradation and restore the activity of glucocerebrosidase in Gaucher disease. Proc Natl Acad Sci U S A 108:21200–21205

23. Moresi V, Williams AH, Meadows E, Flynn JM, Potthoff MJ, McAnally J, Shelton JM, Backs J, Klein WH, Richardson JA, Bassel-Duby R, Olson EN (2010) Myogenin and class II HDACs control neurogenic muscle atrophy by inducing E3 ubiquitin ligases. Cell 143:35–45

24. Pipalia NH, Cosner CC, Huang A, Chatterjee A, Bourbon P, Farley N, Helquist P, Wiest O, Maxfield FR (2011) Histone deacetylase inhibitor treatment dramatically reduces cholesterol accumulation in Niemann-Pick type C1 mutant human fibroblasts. Proc Natl Acad Sci U S A 108:5620–5625

25. Coppede F (2014) The potential of epigenetic therapies in neurodegenerative diseases. Front Genet 5:220

26. Bouchecareilh M, Hutt DM, Szajner P, Flotte TR, Balch WE (2012) Histone deacetylase inhibitor (HDACi) suberoylanilide hydroxamic acid (SAHA)-mediated correction of alpha1-antitrypsin deficiency. J Biol Chem 287:38265–38278

27. Hebert DN, Lamriben L, Powers ET, Kelly JW (2014) The intrinsic and extrinsic effects of N-linked glycans on glycoproteostasis. Nat Chem Biol 10:902–910

28. Gettins PG (2002) Serpin structure, mechanism, and function. Chem Rev 102:4751–4804

29. Hidvegi T, Ewing M, Hale P, Dippold C, Beckett C, Kemp C, Maurice N, Mukherjee A, Goldbach C, Watkins S, Michalopoulos G, Perlmutter DH (2010) An autophagy-enhancing drug promotes degradation of mutant alpha1-antitrypsin Z and reduces hepatic fibrosis. Science 329:229–232

Chapter 19

Expression and Purification of Active Recombinant Human Alpha-1 Antitrypsin (AAT) from *Escherichia coli*

Beena Krishnan, Lizbeth Hedstrom, Daniel N. Hebert, Lila M. Gierasch, and Anne Gershenson

Abstract

Well-established genetic manipulation procedures along with a fast doubling time, the ability to grow in inexpensive media, and easy scaleup make *Escherichia coli* (*E. coli*) a preferred recombinant protein expression platform. Human alpha-1 antitrypsin (AAT) and other serpins are easily expressed in *E. coli* despite their metastability and complicated topology. Serpins can be produced as soluble proteins or aggregates in inclusion bodies, and both forms can be purified to homogeneity. In this chapter, we describe an ion-exchange chromatography-based protocol that we have developed involving the use of two anion-exchange columns to purify untagged human AAT from *E. coli*. We also outline methods that can be used to determine the inhibitory activity of both AAT in cell lysates and purified AAT. Our protocol for the purification of bacterially expressed AAT yields pure and active protein at 6–7 mg/l culture.

Key words Alpha-1 antitrypsin (AAT), Elastase, Trypsin, Anion-exchange chromatography, Stoichiometry of inhibition, Absorbance spectroscopy

1 Introduction

Serine protease inhibitors (serpins) are a superfamily of proteins with diverse sequences and a common two-domain structure with three β-sheets (A, B, and C), 8–9 α-helices, and a loop called the reactive center loop (RCL) connecting sheets A and C [1]. Most serpins are protease inhibitors, and for these serpins, a solvent-exposed RCL is central to the complex mechanism that is used to trap and inactivate target proteases [2–4]. Mature human alpha-1 antitrypsin (AAT, 394 residues) is a prototypical inhibitory plasma serpin that protects lungs during inflammation by irreversibly inhibiting serine proteases released by leukocytes, particularly human leukocyte elastase [5]. Specific single-point mutations in AAT (and other serpins) are known to result in misfolding-associated genetic disorders in humans, which are referred to as the serpinopathies [6].

Florie Borel and Christian Mueller (eds.), *Alpha-1 Antitrypsin Deficiency: Methods and Protocols*, Methods in Molecular Biology, vol. 1639, DOI 10.1007/978-1-4939-7163-3_19, © Springer Science+Business Media LLC 2017

Studies on AAT have contributed significantly to the current understanding of serpins. Investigations of the AAT structure, folding, and/or function using a bottom-up approach require significant amounts of the pure protein. AAT has been produced successfully from both prokaryotic and eukaryotic expression systems. Furthermore, plasma AAT is a glycosylated protein with three N-linked glycans, but glycosylation does not affect its structure, native state stability, or function as a protease inhibitor [7, 8]. Therefore, the Gram negative bacterium *Escherichia coli* (*E. coli*) remains the most robust and economical host for producing recombinant AAT, even though the protein produced in *E. coli* is not glycosylated. Previous reports about the use of *E. coli* for the production and purification of AAT (and its variants) are summarized in Table 1.

Depending on the cell culture growth and protein expression conditions, overexpression of AAT in *E. coli* is known to result in varying extents of protein partitioning between soluble and insoluble (inclusion bodies) forms. Since inclusion bodies generally contain few *E. coli* proteins and are relatively pure, AAT can be expressed at elevated temperatures (such as at 40 °C), which favors the inclusion body formation. Active AAT is then purified from the refolded denaturant-solubilized inclusions bodies (Table 1). However, protein purification from inclusion bodies is more expensive and time consuming than purifying soluble AAT from the *E. coli* cytoplasm, mostly due to the inefficient refolding step. Earlier reports on AAT purification from the soluble cell fraction required either tagged AAT or a purification protocol with more than two steps (Table 1).

Here, we describe a simple two-step anion-exchange chromatography-based protocol for purifying untagged AAT from the cell lysate soluble fraction that can be completed within a day. In order to produce soluble AAT, recombinant AAT expression in *E. coli* is carried out at 30 °C. The first step in the purification process involves running the crude cell lysate through a manually packed column of Q Sepharose Fast Flow resin at pH 5.92, which yields AAT that is approximately 70% pure. In a subsequent step, more than 95% protein purity is achieved by running the eluted sample from the first column through a MonoQ column at pH 8.0.

The mechanism by which active AAT inhibits proteases is through the distortion of the protease active site and this leads to the formation of a relatively stable acyl-enzyme complex [3, 20]. AAT inhibitory activity can be monitored by using sodium dodecyl sulfate polyacrylamide gel electrophoresis (SDS-PAGE) to detect the AAT-protease covalent complex or spectroscopically by monitoring the residual protease activity. While the SDS-PAGE analysis enables detection of various AAT forms (within a complex, free, or cleaved), the spectroscopy-based measurement allows detection and quantification of the extent of protease inhibition. We,

Table 1
Production of AAT protein from *E. coli*

Plasmid; Promoter	Expression Strain	Growth condition; Inducer	Induction Temp.; Time; Yield	Ref.
AAT expression and purification from soluble cell lysate				
pOTSα (with N-term deleted AAT); λ$_{PL}$	AR120	32 °C; Thermal induction at A$_{650}$ of 1.0	42 °C; 60–90 m; NR	[9]
λ$_{PL}$	TGE7213	15 l culture at 30 °C; Thermal induction at A$_{600}$ of 10	42 °C; 6 h; NR	[10]
pTermat; T7	BL21(DE3)	10 l medium 37 °C[a]; IPTG (0.1% w/v) at A$_{600}$ of 10; Rifampicin at 0.1 mg/ml after 30 min of induction	37 °C; 3.5 h; 0.8 mg/g wet cell paste	[11]
pEAT8 (M2 AAT); T7	BL21(DE3)	Semi-defined medium; 0.4 mM IPTG at A$_{600}$ of 1.4	37 °C; 3 h; NR	[12]
pMAL-C2x (MBP-AAT); pTac	BL21CP[b]	0.3 mM IPTG at A$_{600}$ of 0.5[b]	37 °C; 8 h; 7–9 mg/l	[13]
a. pQE31 (His$_6$-AAT); T5 b. pQE30 (His$_6$-AAT); T5	SG13009 (pREP4)	2xYT medium; a. 1.0 mM IPTG at A$_{600}$ of 0.8–1 b. 0.5 – 1.0 mM IPTG at A$_{600}$ of 0.5 – 0.7	a. 30 °C; 3 h; NR b. 37 °C; 4 h; NR	[14] [15]
pEAT8-137 (M2 AAT); T7	BL21(DE3)	37 °C, LB medium; 0.4 mM IPTG at A$_{600}$ of 0.7–0.8	30 °C; 5 h; 6–7 mg/l	[16][c]
AAT expression and purification from inclusion bodies				
pαBcl (N-terminal deleted AAT); λ$_{PL}$	AR120	32 °C until A$_{650}$ of 1.0; Thermal induction	42 °C; 60–90 m; NR	[9]
pEAT8 (M2 AAT); T7	BL21(DE3)	M9ZB medium; 0.4 mM IPTG at A$_{600}$ of 0.8	40 °C; 3 h; NR	[17]
pTermat (Chimeric- AAT); T7	BL21(DE3)	37 °C; IPTG (0.1% w/v) at A$_{600}$ of 10; Rifampicin at 0.1 mg/ml after 30 m of induction	37 °C; 3.5 h; 10–15 mg/l	[18]
pQE30(His$_6$-Multi9AAT); T5	SG-13009	M9 Minimal medium; 1 mM IPTG at A$_{600}$ of 0.5	4.5 h; NR	[19]
pTermat; T7	BL21(DE3)	2xYT medium; 0.5–1.0 mM IPTG at A$_{600}$ of 0.5–0.7; Rifampicin at 0.05 mg/ml after 30 m of induction	37 °C; 4 h; NR	[15]

[a] Defined medium for fermentation
[b] Optimized strain and growth condition from [13]
[c] Current Study
NR not reported

therefore, provide protocols for measuring AAT activity using both methods: (a) the inhibition of porcine pancreatic elastase monitored by SDS-PAGE and (b) bovine pancreatic trypsin inhibition using an absorbance-based assay. The mobility shift of the AAT-protease complex as determined by gel electrophoresis is also the basis of our activity screen to test for functional AAT in soluble cell lysates prior to purification. Such an assay is useful to screen for the effects of mutations on AAT activity as long as some fraction of the variant protein is expressed as soluble protein. This screen can also guide subsequent purification.

2   Materials

2.1   AAT Expression Plasmid

Plasmid pEAT8-137—The parent plasmid, pEAT8, harbors a naturally occurring variant of human AAT, the M2 variant containing the Arg101His, and Glu376Asp mutations relative to the canonical mature protein sequence [21, 22]. M2 was cloned under the T7 promoter in pET8c [17]. A synonymous codon substitution at position 137, TTG(Leu) to CTG(Leu), was introduced in order to eliminate an internal translation initiation site that is downstream of an internal Shine-Dalgarno site. Two additional synonymous substitutions at nucleotide positions 399 and 402 (T399C and T402C) were made to introduce a unique AatII restriction site in order to confirm the mutations through restriction mapping, and the resulting plasmid is pEAT8-137 [23]. The pEAT8-137 plasmid is used as an AAT protein expression vector (*see* **Note 1**).

2.2   Cell Growth and Protein Expression

1. *E. coli* strain BL21(DE3).
2. Luria Bertaini (LB) broth.
3. Ampicillin.
4. Isopropyl β-D-1-thiogalactopyranoside (IPTG).

2.3   Screening for AAT Activity in Cell Lysates Using Western Blots

1. Lysis buffer: 50 mM HEPES sodium salt, 100 mM NaCl, 10 mM $CaCl_2$, pH 8.0.
2. Porcine pancreatic elastase.
3. Bovine pancreatic trypsin.
4. 1 mM and 0.1 M Hydrochloric acid (HCl).
5. Sodium dodecyl sulfate (SDS) sample buffer with freshly added β-mercaptoethanol.
6. 12% Tris-Glycine SDS-PAGE.
7. Polyvinylidene difluoride (PVDF) membrane (Biotrace Pall Life Sciences, 0.45 μm).
8. Transfer buffer: 25 mM Tris, pH 8.3, 192 mM Glycine, 15% (v/v) Methanol, 0.05% (w/v) SDS.

9. 1× Tris Buffered Saline (TBST): 50 mM Tris, 0.9% (w/v) NaCl pH adjusted to 7.6 with HCl, 0.05% (v/v) Tween-20.

10. Antibody against AAT (Rabbit Polyclonal AAT antibody, TA590016, Origene, *see* **Note 2**).

11. Secondary antibody against the primary antibody (HRP-conjugated-Goat anti-Rabbit secondary antibody, Pierce).

12. Western blot detection substrate (Clarity, Bio-Rad).

13. PVDF membrane staining solution: 0.5% (w/v) Naphthol blue black (Amido black, Sigma), 25% (v/v) isopropanol, 10% (v/v) acetic acid.

14. PVDF membrane destaining solution: 20% (v/v) methanol, 7.5% (v/v) acetic acid.

2.4 AAT Protein Purification

1. Resuspension buffer: 10 mM Bis-Tris, pH 5.92.

2. 2.5 cm diameter glass column with appropriate flow adaptor.

3. Q Sepharose Fast Flow resin.

4. Buffers for running sample on Q Sepharose Fast Flow column:
   Buffer A: 10 mM Bis-Tris, pH 5.92.
   Buffer B: 10 mM Bis-Tris, 1 M NaCl, pH 5.92.

5. MonoQ 4.6/100 PE (GE Healthcare, *see* **Note 3**).

6. Buffers for running sample on MonoQ 4.6/100 PE column:
   Buffer C: 10 mM HEPES sodium salt, pH 8.0.
   Buffer D: 10 mM HEPES sodium salt, 1 M NaCl, pH 8.0.

7. 30 kDa MWCO protein concentrator (Amicon Ultra 15 ml centrifugal filter).

2.5 Gel-Based Stoichiometry of Inhibition (SI) Determination

1. Purified AAT.

2. Assay buffer: 50 mM HEPES sodium salt, 100 mM NaCl, 10 mM $CaCl_2$, pH 8.0.

3. Porcine pancreatic elastase.

4. Bovine pancreatic trypsin.

5. 1 mM and 0.1 M Hydrochloric acid (HCl).

6. Sodium dodecyl sulfate (SDS) sample buffer with freshly added β-mercaptoethanol.

7. 12% Tris-Glycine SDS-PAGE.

2.6 Trypsin Active Site Titration

1. 4-nitrophenyl-4-guanidinobenzoate hydrochloride (pNGB).

2. Dimethylformamide (DMF).

3. Assay buffer: 50 mM HEPES sodium salt, 100 mM NaCl, 10 mM $CaCl_2$, pH 8.0.

4. Bovine pancreatic trypsin.

5. 1 mM Hydrochloric acid (HCl).

2.7 Spectroscopy-Based SI Determination

1. Nα-Benzoyl-L-arginine 4-nitroanilide hydrochloride (BAPNA).

2. Dimethylformamide (DMF).

3. Assay buffer: 50 mM HEPES sodium salt, 100 mM NaCl, 10 mM $CaCl_2$, pH 8.0.

4. Bovine pancreatic trypsin.

5. Purified AAT.

2.8 Special Equipment

1. Bacterial cell lysis equipment (Sonicator, Sonics Vibra-cell model VCX 500 or Microfluidizer processor, model M-110L, Microfluidics, Newton MA).

2. Western blot electroblotting unit (Mini Trans-Blot cell, Bio-Rad).

3. Chemiluminescence imaging system (GBOX Chemi-XRQ gel documentation unit, Syngene).

4. FPLC protein purification system (ÅKTA Purifier, GE Healthcare).

5. UV-Vis spectrophotometer (Varian).

3   Methods

3.1 Cell Growth and AAT Protein Expression

1. Transform *E. coli* BL21(DE3) with the pEAT8–137 plasmid.

2. Inoculate 3–50 ml of LB that contains ampicillin at a 100 μg/ml final concentration with BL21(DE3) cells freshly transformed with the pEAT8-137 plasmid and grow the culture overnight at 30 °C or 37 °C, shaking at 200–250 rpm.

3. Typically, culture volumes of 10 ml are used to screen variants for AAT activity in cell lysates, and 2 l cultures are grown for protein purification experiments. In particular, 1–2% of the overnight grown culture is used to inoculate fresh LB that contains ampicillin (100 μg/ml), and cells are grown at 37 °C until an OD_{600} of approximately 0.7–0.8 is reached. For larger overnight cultures (e.g., 50 ml), spin down and resuspend the cells in fresh LB medium that contains ampicillin (100 μg/ml) since these cells will have used up the ampicillin.

4. IPTG should then be added to a final concentration of 0.4 mM in order to induce AAT protein expression and the bacteria are grown with shaking at 30 °C for 5 h (*see* **Note 4**).

5. Cells are harvested at 4 °C by centrifugation at 4000 × *g* for 10 min. The cell pellet from a small scale growth (10 ml culture) is resuspended in 0.5 ml of lysis buffer (*see* **item 1** in

Subheading 2.3) and 1 l of culture is resuspended in about 40 ml of resuspension buffer (*see* **item 1** in Subheading 2.4). The resuspended cell suspension should be stored at −80 °C for later use.

3.2 Screening for AAT Activity in Cell Lysates Using Western Blots

1. The frozen cell pellet suspension is thawed and lysed by sonication on ice using Sonics Vibra-cell model VCX 500 with a 3 mm microtip probe at 30% amplitude for 10 s, with 15 s off time and about 10–11 pulses.

2. The lysate is centrifuged at 4 °C for 10 min at 12,000 × *g* and the soluble fraction (supernatant) that is separated from the cell debris is kept on ice or at 4 °C.

3. A 2 mg/ml stock solution of both porcine pancreatic elastase (in water) and bovine pancreatic trypsin (in 1 mM HCl) should be prepared.

4. Protease inhibition activity assays are set up in a reaction volume of 15 μl with 1 μl protease (elastase or trypsin), 9 μl of assay buffer, and 5 μl of the cell lysate (supernatant fraction) at 25 °C for 10 min. In order to avoid acidifying the reaction, the protease is added to the assay buffer before the AAT-containing cell lysate is added. The reactions are quenched by adding 1 μl of 0.1 M HCl and SDS-PAGE sample buffer that contains β-mercaptoethanol (1×), and then the samples are boiled for 3–5 min. An assay with purified AAT serves as a positive control (*see* **Note 5**).

5. Samples are analyzed on a 12% Tris-Glycine SDS-PAGE gel and the resolved proteins are transferred onto a PVDF membrane using the Bio-Rad Mini Trans-Blot cell. The transfer is carried out for 1 h at a setting of 100 V and constant 350 mA current.

6. The membrane is blocked in 5% (w/v) of Bio-Rad blotting grade blocker solution in 1× TBST for 1 h at room temperature (or overnight at 4 °C).

7. The membrane is incubated with the diluted primary antibody against AAT for 1 h (with gentle shaking) followed by an incubation with the diluted secondary antibody for 1 h at room temperature. The antibody dilutions are made in 1× TBST following the manufacturer's recommendations. The membrane is washed with agitated shaking at least three times for 5 min each with 1× TBST between the each antibody incubation.

8. The membrane is probed using the Clarity Western ECL blotting substrate following the manufacturer's protocol and digitally imaged on the Syngene Chemidoc unit (*see* Fig. 1a).

9. After developing the blot, the membrane is irreversibly stained with the amido black PVDF membrane staining solution for less than 5 min followed by extensive destaining in the PVDF membrane destaining solution (*see* Fig. 1b).

202   Beena Krishnan et al.

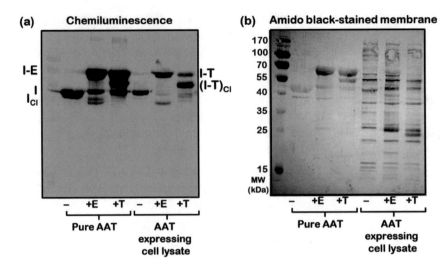

Fig. 1 Screening AAT protease inhibition activity in cell lysates. AAT protein in the soluble fraction of cell lysates is tested for function by monitoring the formation of a covalent complex between AAT [I] and proteases, elastase [E] and trypsin [T]. (**a**) The inhibitor-protease complexes [I-E and I-T] are detected by chemiluminescence imaging of the Western blot by using a primary antibody against AAT. (**b**) The sample and protein transfer quality are assessed by staining the membrane with Amido black. In the case of the AAT interaction with trypsin, degradation products of the complex, [I-T]$_{Cl}$, are observed [3]. Cleaved AAT [I$_{Cl}$] is visible in reactions between AAT and elastase

3.3 AAT Protein Purification

3.3.1 Cell Lysis

1. Thaw the frozen cell pellet under cold water.

2. Cell lysis is carried out on ice using Sonics Vibra-cell model VCX 500 with a 13 mm diameter probe with a replaceable tip at 30% amplitude for 10 s, with 15 s off time intervals and about 40 pulses. For an 80 ml cell suspension, the total sonication time is 30 min). Alternatively, cell lysis may be carried out on a Microfluidizer processor at a pressure of 16–17 k PSI with 4 passages through the system (*see* **Note 6**).

3. The cell lysate is spun at 18,000 × g, 4 °C for 45 min in order to obtain the soluble fraction (supernatant). The supernatant is filtered using a 0.45 μm filter prior to loading it onto the Q Sepharose Fast Flow column. Protein purification is carried out following a 2-step (or column) purification protocol, as described below.

3.3.2 First Chromatographic Step of the AAT Protein Purification—Soluble Cell Extract on a Manually Packed Q Sepharose Fast Flow Column (Column 1)

1. Pack the Q Sepharose Fast Flow resin in a 2.5 cm diameter glass column in order to achieve a final bed volume of 20 ml for processing the supernatant from a 2 l growth culture.

2. Filtered supernatant is loaded onto the Q Sepharose column that is equilibrated with buffer A at 4 °C.

3. Protein is eluted with a linear gradient of 50–140 mM NaCl in buffer A (e.g., 5–14% Buffer B) in 9 column volumes

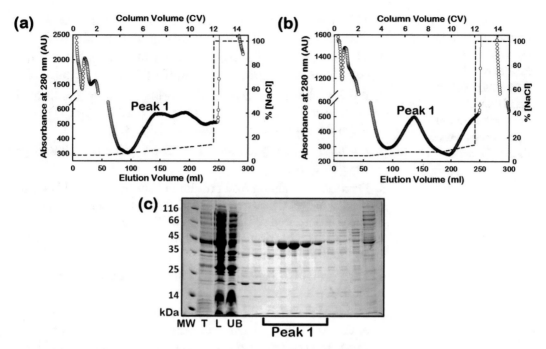

Fig. 2 Step 1 of the AAT protein purification process. FPLC chromatogram of AAT in the supernatant (labeled "L") eluted from the Q Sepharose Fast Flow column using (**a**) a continuous linear salt gradient method, or (**b**) a 3-step gradient method. SDS-PAGE of fractions eluted from column 1 in (**b**) is shown in panel (**c**). The fractions within peak 1 contain AAT. The *dashed line* shows the salt gradient used for elution. The labels, T and UB, denote the total cell lysate and flow through fraction from the column respectively

(CV) (*see* Fig. 2a) or alternatively using a 3-step elution method with a linear gradient of 50–80 mM NaCl in Buffer A (5–8% Buffer B, 3 CV), and a step gradient at 80 mM NaCl in Buffer A (8% Buffer B, 3 CV), followed by a linear gradient of 80–140 mM NaCl in Buffer A (8–14% Buffer B, 3 CV) (*see* Fig. 2b).

4. The eluted protein fractions are analyzed on a 12% Tris-Glycine SDS-PAGE gel (*see* Fig. 2c).

5. AAT elutes from the column between 70 and 100 mM NaCl (e.g., 7–10% buffer B) in a linear gradient elution (*see* peak 1 in Fig. 2a), or within the 80 mM NaCl elution step (e.g., 8% Buffer B) in the alternative step gradient elution method (*see* peak 1 in Fig. 2b).

6. Fractions containing AAT (peak 1) are pooled and concentrated using a 30 kDa MWCO protein concentrator at 4 °C and buffer exchanged into Buffer C (*see* **Note 7**).

3.3.3 Second
Chromatographic Step
of the AAT Protein
Purification—Final
Purification of AAT Protein
Eluted from the
Q Sepharose Fast Flow
Column Using a MonoQ
Column (Column 2)

1. The concentrated protein sample from column 1 is filtered through a 0.2 μm filter to remove any aggregated protein.

2. The filtered sample is applied to the MonoQ column that is equilibrated in 70 mM NaCl in Buffer C (e.g., in 7% Buffer D).

3. Protein elution is carried out using a linear gradient of 70–200 mM NaCl in buffer C (e.g., 7–20% buffer D) in 50 CV and 1 ml fractions are collected (Fig. 3a).

4. Fractions are analyzed on a 12% Tris-Glycine SDS-PAGE gel (*see* Fig. 3b).

5. Fractions containing AAT are eluted between 120 and 160 mM NaCl as a broad peak (*see* peak 2 in Fig. 3a, b), and then pooled and concentrated using a 30 kDa MWCO protein concentrator at 4 °C. The fraction is then buffer exchanged into the desired final buffer (Buffer C).

6. Pure protein is flash frozen in liquid nitrogen and stored at −80 °C.

7. The final protein yield is about 6–7 mg/l culture for wild-type AAT, with the protein concentration measured using an extinction coefficient at 280 nm ($E_{1cm, 280 nm}(1\%)$) of 5.3 [24].

8. The extent of purification achieved after each purification step described above is estimated from sample analyses on a 12% Tris-Glycine SDS-PAGE gel (*see* Fig. 3c).

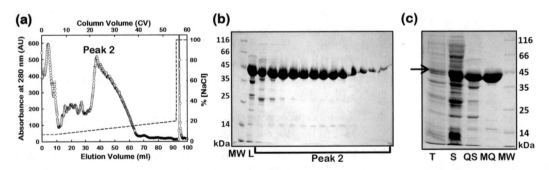

Fig. 3 Step 2 of the AAT protein purification process. FPLC chromatogram for the MonoQ column following application of peak 1 (labeled "L") from the Q Sepharose purification step (**a**). Fractions eluted from the MonoQ column are analyzed using SDS-PAGE (**b**). Peak 2 fractions contain the final pure protein. The salt gradient for the elution is indicated by the *dashed line*. An overall view of protein purity achieved in the individual purification steps can be assessed from panel (**c**). Samples from different steps of the purification protocol; total cell lysate [T], soluble fraction of the cell lysate [S], eluted protein from the Q Sepharose column [QS], and final pure AAT from the MonoQ column [MQ], are analyzed on a 12% SDS-PAGE gel. The protein is >95% pure after the final column. The *arrow* in panel C marks AAT in the cell lysate

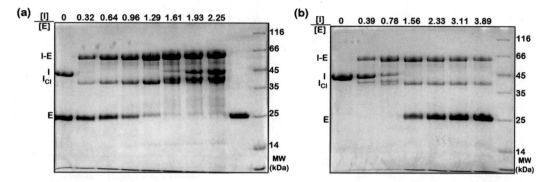

Fig. 4 Stoichiometry of inhibition (SI) of elastase by AAT. SDS-PAGE analyses of AAT-elastase covalent complex [I-E] formation in reaction mixtures containing: (**a**) an increasing concentration of AAT [I] with a fixed concentration of elastase [E], and (**b**) an increasing concentration of elastase [E] with a fixed concentration of AAT [I]. The SI of elastase by AAT is estimated as 1.6 in this measurement, which is in agreement with the previous reports [25–27]. I_{cl} indicates cleaved AAT

3.4 Gel-Based Stoichiometry of Inhibition (SI) Determination

1. Prepare stock solution of porcine pancreatic elastase (2 mg/ml in water).

2. The AAT activity assay is set up in a reaction volume of 15 μl that contains purified AAT protein and elastase. Molar ratios of AAT to elastase from 0 to 2.0 are used by keeping one of the protein concentrations fixed and increasing the concentration of the other protein (*see* Fig. 4). The samples are incubated at 25 °C for 10 min.

3. The reactions are quenched by adding 1 μl of 0.1 M HCl and SDS-PAGE sample buffer that contains β-mercaptoethanol (1×), and then boiling the samples for 3 min.

4. Samples are analyzed on a 12% Tris-Glycine SDS-PAGE gel (*see* Fig. 4). The molar ratio at which all of the elastase is consumed and is present only in covalent AAT-elastase (I-E, *see* Fig. 4) complexes (*see* in Fig. 4a) or all of the AAT is present as I-E (*see* in Fig. 4b) represents the SI of elastase by AAT. A similar gel-based assay may be performed to determine the SI of trypsin by AAT (data not shown).

3.5 Trypsin Active Site Titration

The active site titration is used to measure the concentration of active protease (*see* **Note 8**). One molecule of active trypsin hydrolyses one molecule of *p*-nitrophenyl-*p*'-guanidino benzoate (pNGB) resulting in the stoichiometric release of *p*-nitrophenol and an inactive trypsin molecule [28]. The *p*-nitrophenol molecule absorbs light at 412 nm. The concentration of the active enzyme is determined by measuring the burst of absorbance at 412 nm when this reaction takes place.

1. Stock solutions of 0.1 M pNGB in DMF and approximately 2 mg/ml of trypsin in 1 mM HCl should be prepared.

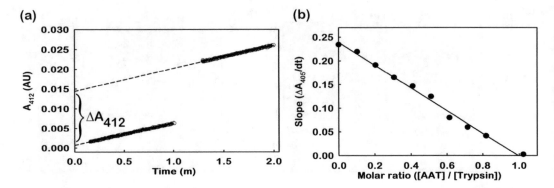

Fig. 5 Inhibition of trypsin by AAT. (**a**) Trypsin active site titration using pNGB. Hydrolysis of pNGB by trypsin releases *p*-nitrophenol and an increase in the absorbance is observed at 412 nm (A_{412}). The change in absorbance (ΔA_{412}) is used to calculate the concentration of active trypsin. The dotted line represents a linear fit to the data. (**b**) The stoichiometric inhibition of trypsin by AAT is measured by monitoring the residual trypsin activity in reactions that contain increasing molar ratios of AAT to trypsin and by using BAPNA, a trypsin substrate. The line in the plot is a linear fit to the data (*filled circles*). The x-intercept obtained from the fitted line (0.99 in this data figure) is the SI for AAT-trypsin interaction

2. Two μl of pNGB is added to 1 ml of assay buffer and the baseline is monitored at 412 nm for about 1 min. The cuvette is removed without stopping the spectrophotometer and 5–20 μl of the enzyme is added to the solution. The sample is quickly mixed and the cuvette is placed back in the spectrophotometer to measure the absorbance burst by continuing the absorbance measurement at 412 nm (A_{412}) for one additional minute (*see* Fig. 5a).

3. The above measurement is repeated at least three times and an average of the three determinations is used to calculate the concentration of active trypsin.

4. The change in A_{412} before and after adding trypsin is recorded as the ΔA_{412}. The active site concentration, i.e., active trypsin concentration, is calculated as
   $[\text{Trypsin}]_{\text{Active}} = \Delta A_{412}/\varepsilon_{412}$, where $\varepsilon_{412} = 17{,}000 \text{ M}^{-1}\text{cm}^{-1}$
   at pH 8.0 for the *p*-nitrophenol released from pNGB.

3.6 Spectroscopy-Based SI Determination

BAPNA is a chromogenic trypsin substrate [29]. Trypsin cleaves BAPNA, thereby releasing *p*-nitroaniline that absorbs around 400 nm. In a reaction with trypsin and BAPNA, the increase in absorbance at 405 nm in relation to the time is proportional to trypsin activity. The incubation of trypsin with AAT leads to the formation of an AAT-trypsin complex, thereby decreasing the concentration of free, active trypsin. Therefore, measuring the residual active trypsin concentration in reaction mixtures that contain a preformed AAT-trypsin complex using BAPNA provides an estimation of the stoichiometric inhibition of trypsin by AAT.

1. A 100 μl stock solution of 0.1 M BAPNA in DMF should be prepared (*see* **Note 9**).

2. The trypsin stock solution with the active protease concentration predetermined (section 3.5) should be used for the SI determination.

3. Mix the AAT and trypsin at molar ratios ranging from 0 to 1.2 (or 2.0 if assaying new variants) in 1.5 ml centrifuge tubes, making sure to keep the trypsin concentration fixed while increasing the AAT concentration. The total reaction volume is 100 μl. To avoid acidifying the AAT, trypsin is added to the assay buffer and mixed before the addition of AAT. The samples are incubated at 25 °C for 15 min.

4. After 15 min, add assay buffer to bring the volume to 1 ml.

5. Monitor the residual trypsin activity by adding 5 μl of BAPNA and measuring the absorbance at 405 nm (A_{405}) for about 2 min.

6. The plots of A_{405} versus time should be fitted to a line and the slope ($\Delta A_{405}/dt$) should be recorded.

7. Next, a plot of $\Delta A_{405}/dt$ versus molar ratio of AAT to trypsin ($[AAT]/[Trypsin]_{Active}$) should be made. The x-intercept obtained from the linear fit of the data yields the SI of trypsin by AAT (*see* Fig. 5b). An SI value of 1 indicates that one mole of AAT inhibits one mole of trypsin and that the purified AAT is 100 percent active.

---

4   Notes

1. The protocols described here have also been used successfully for AAT variant genes cloned in pET29b(+) and pET16b plasmids. For AAT variants that contain exposed cysteine residues, 1 mM β-mercaptoethanol is added to all of the buffers that are used for purification just before use. All buffers should be filtered through a 0.2 μm membrane.

2. The monoclonal antibody against AAT from Abcam (ab9399) also detects AAT-protease complexes in cell lysates [16]. The primary antibody against the protease may also be used in this assay, and we have tested an anti-elastase antibody in our previous study [16].

3. A smaller MonoQ column, MonoQ 5/50 Global (GE Healthcare), has also been used for the second purification step.

4. Studies of temperature-dependent wild-type AAT expression at 30 °C, 37 °C and 40 °C indicate that the maximal amount of soluble protein is produced at 30 °C (post-IPTG addition). For purification from inclusion bodies, 40 °C may be used.

5. Unlike trypsin, excess elastase does not digest the AAT-elastase complex. Therefore, an excess of elastase can be used in the assay.

6. Lysozyme from chicken egg white (Sigma) and DNaseI from bovine pancreas, grade II (Roche) are added to the thawed cell suspension to final concentrations of 6.25 μg/ml and 2.5 μg/ml, respectively, and incubated on ice for 30 min prior to cell lysis by sonication.

7. Alternately, the pooled fractions from the Q-Sepharose column may be dialyzed against 10 mM HEPES sodium salt, pH 8 overnight at 4 °C, and then concentrated before loading it on the MonoQ column.

8. The fraction of active trypsin can be determined from the ratio of the concentration of active trypsin ($[\text{Trypsin}]_{\text{Active}}$) to that of the concentration of prepared trypsin stock solution: ($[\text{Trypsin}]_{\text{Total}}$), $[\text{Trypsin}]_{\text{Active}}/[\text{Trypsin}]_{\text{Total}}$.

9. The BAPNA stock solution must be fresh. BAPNA is very sensitive to hydrolysis and should be kept dry. When making BAPNA stock solutions, take the BAPNA out of the freezer and do not open the bottle until it is at room temperature. Once the BAPNA is at room temperature, open the bottle and weigh out the appropriate amount. Immediately return the bottle to the freezer. If the absorbance at 405 nm versus the time plots are curved (not linear), then the BAPNA must be replaced.

Acknowledgment

We thank Prof. C. L. Cooney (Massachusetts Institute of Technology) for providing the pEAT8-137 plasmid, and Prof. M. H. Yu (Korea Institute of Science and Technology), who provided the original source of the pEAT8 plasmid. This work was supported by grants from the Alpha-1 Foundation and DBT-Ramalingaswami Re-entry Fellowship (to B.K.), the US National Institutes of Health, OD-00045 (to L.M.G.), and R01 GM094848 (to A.G., D.N.H. & L.M.G.). B.K. thanks CSIR-IMTECH for the protein instrumentation facility.

References

1. Gettins PG (2002) Serpin structure, mechanism, and function. Chem Rev 102:4751–4804

2. Huntington JA, Carrell RW (2001) The serpins: nature's molecular mousetraps. Sci Prog 84:125–136

3. Huntington JA, Read RJ, Carrell RW (2000) Structure of a serpin-protease complex shows inhibition by deformation. Nature 407:923–926

4. Stratikos E, Gettins PG (1999) Formation of the covalents erpin-proteinase complex

involves translocation of the proteinase by more than 70 Å and full insertion of the reactive center loop into beta-sheet A. Proc Natl Acad Sci U S A 96:4808–4813

5. Stockley RA (2014) Alpha1-antitrypsin review. Clin Chest Med 35:39–50

6. Gooptu B, Lomas DA (2009) Conformational pathology of the serpins: themes, variations, and therapeutic strategies. Annu Rev Biochem 78:147–176

7. Powell LM, Pain RH (1992) Effects of glycosylation on the folding and stability of human, recombinant and cleaved α1-antitrypsin. J Mol Biol 224:241–252

8. Sarkar A, Wintrode PL (2011) Effects of glycosylation on the stability and flexibility of a metastable protein: the human serpin α1-antitrypsin. Int J Mass Spectrom 302:69–75

9. Johansen H, Sutiphong J, Sathe G, Jacobs P, Cravado, A, Bollen A, Rosenberg M, Shatzman A (1987) High-level production of fully active human α1-antitrypsin in Escherichia coli, Mol Biol & Med 4:291–305.

10. Bischoff R, Speck D, Lepage P, Delatre L, Ledoux C, Brown SW, Roitsch C (1991) Purification and biochemical characterization of recombinant α1-antitrypsin variants expressed in *Escherichia coli*. Biochemistry 30:3464–3472

11. Hopkins PC, Carrell RW, Stone SR (1993) Effects of mutations in the hinge region of serpins. Biochemistry 32:7650–7657

12. Griffiths SW, Cooney CL (2002) Development of a peptide mapping procedure to identify and quantify methionine oxidation in recombinant human α1-antitrypsin. J Chromat A 942:133–143

13. Agarwal S, Jha S, Sanyal I, Amla DV (2010) Expression and purification of recombinant human α1-proteinase inhibitor and its single amino acid substituted variants in *Escherichia coli* for enhanced stability and biological activity. J Biotech 147:64–72

14. Zhou A, Carrell RW, Huntington JA (2001) The serpin inhibitory mechanism is critically dependent on the length of the reactive center loop. J Biol Chem 276:27541–27547

15. Pearce MC, Cabrita LD (2011) Production of recombinant serpins in *Escherichia coli*. Methods Enzymol 501:13–28

16. Krishnan B, Gierasch LM (2011) Dynamic local unfolding in the serpin α-1 antitrypsin provides a mechanism for loop insertion and polymerization. Nat Struct Mol Biol 18:222–226

17. Kwon KS, Lee S, Yu MH (1995) Refolding of α1-antitrypsin expressed as inclusion bodies in *Escherichia coli*: characterization of aggregation. Biochim Biophys Acta 1247:179–184

18. Bottomley SP, Stone SR (1998) Protein engineering of chimeric Serpins: an investigation into effects of the serpin scaffold and reactive centre loop length. Protein Eng 11:1243–1247

19. Peterson FC, Gordon NC, Gettins PG (2000) Formation of a noncovalent serpin-proteinase complex involves no conformational change in the serpin. Use of ^1H-^{15}NHSQC NMR as a sensitive nonperturbing monitor of conformation. Biochemistry 39:11884–11892

20. Dementiev A, Dobo J, Gettins PG (2006) Active site distortion is sufficient for proteinase inhibition by serpins: structure of the covalent complex of α1-proteinase inhibitor with porcine pancreatic elastase. J Biol Chem 281:3452–3457

21. Huber R, Carrell RW (1989) Implications of the three-dimensional structure of α1-antitrypsin for structure and function of serpins. Biochemistry 28:8951–8966

22. UniProt Consortium (2015) UniProt: a hub for protein information. Nucleic Acids Res 43:D204–D212

23. Laska, M. E. (2001) The effect of dissolved oxygen on recombinant protein degradation in *Escherichia coli*, Thesis, Massachusetts Institute of Technology, Cambridge, MA

24. Pannell R, Johnson D, Travis J (1974) Isolation and properties of human plasma α-1-proteinase inhibitor. Biochemistry 13:5439–5445

25. James HL, Cohen AB (1978) Mechanism of inhibition of porcine elastase by human alpha-1-antitrypsin. J Clin Invest 62:1344–1353

26. Seo EJ, Im H, Maeng JS, Kim KE, Yu MH (2000) Distribution of the native strain in human α1-antitrypsin and its association with protease inhibitor function. J Biol Chem 275:16904–16909

27. Dolmer K, Gettins PG (2012) How the serpin α1-proteinase inhibitor folds. J Biol Chem 287:12425–12432

28. Chase T Jr, Shaw E (1967) P-Nitrophenyl-p'-guanidinobenzoate HCl: a new active site titrant for trypsin. Biochem Biophys Res Comm 29:508–514

29. Erlanger BF, Kokowsky N, Cohen W (1961) The preparation and properties of two new chromogenic substrates of trypsin. Arch Biochem Biophys 95:271–278

Chapter 20

Quantification of Total Human Alpha-1 Antitrypsin by Sandwich ELISA

Qiushi Tang, Alisha M. Gruntman, and Terence R. Flotte

Abstract

In this chapter we describe an enzyme-linked immunosorbent assay (ELISA) to quantitatively measure human alpha-1 antitrypsin (AAT) protein levels in serum, other body fluids or liquid media. This assay can be used to measure the expression of the human AAT (hAAT) gene in a variety of gene transfer or gene downregulation experiments.

A hAAT-specific capture antibody and a HRP-conjugated anti-AAT detection antibody are used in this assay. The conjugated anti-AAT used in this protocol, instead of the typical sandwich which employs an unconjugated antibody followed by a specifically conjugated IgG, makes the assay simpler and decreases variability. This provides a useful tool to evaluate the AAT levels in clinical and research samples and can allow fairly rapid testing of a large number of samples.

Key words Alpha-1 antitrypsin, AAT, Enzyme-linked immunosorbent assay, Sandwich ELISA

1 Introduction

The following protocol details the performance of an enzyme-linked immunosorbent assay (ELISA) to quantitate levels of human alpha-1 antitrypsin in both research and clinical samples. The assay is species-specific for human AAT in the background of mouse serum, but will not differentiate between hAAT from endogenous serum AAT in nonhuman primate species [1]. The assay is not allele-specific. Therefore, it is not generally useful for measuring expression of wild-type (PiM) hAAT protein in the serum of patients with AAT deficiency due to the common missense E342K (PiZ) mutation, but could be used to measure total (wild-type + mutant) hAAT levels in that setting.

The human alpha-1 antitrypsin (AAT) "sandwich" ELISA assay relies on an AAT-specific capture antibody which will bind the AAT protein to the surface of the plate, after which any unbound antigen is removed by washes. A skim milk blocking buffer is then used to inhibit any nonspecific binding to the well surface. Samples to be

Florie Borel and Christian Mueller (eds.), *Alpha-1 Antitrypsin Deficiency: Methods and Protocols*, Methods in Molecular Biology, vol. 1639, DOI 10.1007/978-1-4939-7163-3_20, © Springer Science+Business Media LLC 2017

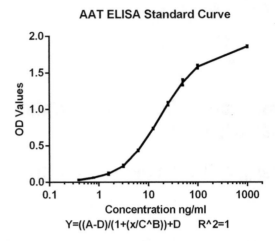

AAT ELISA Standard Curve

$$Y=((A-D)/(1+(x/C^B))+D \qquad R^2=1$$

Fig. 1 Assay sensitivity: The assay was performed as described in the protocol, with increasing amounts of AAT in the protein standard. OD values for the standard are: **1.582**, 1.367, 1.076, 0.734, 0.434, **0.222**, 0.115, and 0.062. As shown here, the linear range of the standard curve is between 1.582 and 0.222, which represents the dynamic range for this assay

assayed are added and human AAT is bound by the capture antibody. Then the detection antibody, which is conjugated to horseradish peroxidase (HRP) (an enzyme leading to a signal readout) is added. 3,3′,5,5′-tetramethylbenzidine (TMB) microwell peroxidase substrate system is used to develop a deep blue color in the presence of peroxidase-labeled conjugates. The density of yellow coloration after addition of the stop solution is directly proportional to the amount of alpha-1 antitrypsin bound in the well. The ELISA is sensitive and has a wide dynamic range (the standard curve linear range is from 1 to 100 ng/ml) (Fig. 1).

Since AAT is a secreted protein, this ELISA can detect AAT in a wide range of samples, including human serum, plasma, cell culture supernatant, and milk as well as in AAT transgenic mouse serum (mice expressing human ZAAT). We have tested the ELISA using serum from multiple species, including several nonhuman primates, and proven that the native AAT produced in those species does not cross-react with the human antibody. Tested species include: rhesus macaques, cynomolgus macaques, African green monkeys, marmosets, ferrets, and wild-type C57/B6 mice. However, the ELISA as described does not differentiate between normal AAT and mutant human AAT.

2   Materials

The reagents should be prepared as described below and stored at room temperature unless otherwise noted. Any material or sample waste should be disposed of according to your institution's waste handling regulations. Millipore water should be used for all reagents unless the protocol specifies otherwise.

1. Clear flat-bottom nonsterile 96-well plate

2. Microplate sealing films, nonsterile.

3. Voller's buffer: Add the follow chemicals: 2.76 g Na_2CO_3 (106.0 g/mol), 1.92 g $NaHCO_3$ (84.01 g/mol), 0.2 g NaN_3 (65.01 g/mol), bring total volume to 1 l using water. The final pH should be adjusted to 9.6 with NaOH.

4. Wash buffer: PBS–Tween: 500 μl Tween 20 added to 1 l of 1× PBS

5. Blocking buffer: 1% skim milk in wash buffer (above). Add 1.0 g of skim milk powder to 100 ml of wash buffer.

6. Coating antibody: Cappel goat antiserum to human alpha-1 antitrypsin (Cat # 55111, Cappel Laboratories, MP Biomedicals, Santa Ana, CA). Make 5 mg/ml stock by reconstituting with 2 ml of dH2O. Make 23 μl aliquots and store at −80 °C. Working concentration: add 22 μl of aliquoted antibody to 11 ml of coating buffer (1:500 dilution).

7. Human AAT standard: Human alpha-1 antitrypsin (Cat# 16-16-011609, Athens Research and Technology, Athens, GA) make the stock solution 1 mg/ml by reconstituted in 1 ml PBS. Make 20 μl aliquots and store at −80 °C. The dilution scheme for the standard is demonstrated in the table (Table 1) below. All dilutions are performed using blocking buffer (*see* above). Add 100 μl of standard from **steps 1–9** to individual wells of the assay plate. It is recommended to run the standard in duplicate (*see* **Note 1**).

Table 1
Standards dilution chart

Step	Final hAAT concentration (ng/ml)	hAAT Standard	Sample diluent w/blocking buffer
0	1000	10 μl of 1 mg/ml stock	10 ml
1	100	100 μl from step 0	900 μl
2	50	500 μl from step 1	500 μl
3	25	500 μl from step 2	500 μl
4	12.5	500 μl from step 3	500 μl
5	6.25	500 μl from step 4	500 μl
6	3.125	500 μl from step 5	500 μl
7	1.5625	500 μl from step 6	500 μl
8	0.78125	500 μl from step 7	500 μl

8. Unknown samples: Optimal sample dilutions are dependent on the condition and should be determined by the user. Samples are diluted to desired concentrations in the standard range with blocking buffer. Duplicate wells are recommended (*see* **Notes 1** and **2**).

9. Positive control: Human serum (Cat#14-102E, BioWhittaker™ Media, Lonza Walkersville, Inc., Walkersville, MD). 20 μl aliquots stored at −80 °C. Working dilution is 1:200,000 diluted in blocking buffer. Duplicate wells (2/plate) are recommended.

10. Secondary antibody: Anti-alpha-1 antitrypsin antibody (HRP conjugated) (Cat#7635, Abcam, Cambridge, UK) should be diluted 1:10 in 1× PBS. Store 23 μl aliquots at −80 °C. Working concentration: 22 μl of aliquoted antibody diluted in 11 ml of blocking buffer (1:500 dilution).

11. TMB peroxidase substrate (TMB microwell peroxidase substrate system, SeraCare Life Sciences) stored at 4 °C: prepare freshly by mixing 6 ml of TMB solution A with 6 ml of TMB solution B per 96-well plate being run.

12. Stop solution: 10 ml 95–98% of sulfuric acid, 90 ml of water.

13. Plate washer (optional) (MultiWash III Microplate Washer, TriContinent, Grass Valley, CA).

14. Microplate reader with 450 nm filter, such as VersaMax (Molecular Devices, Sunnyvale, CA).

15. Software (optional), such as SoftMax Pro (Molecular Devices).

3 Methods

1. Add 100 μl of diluted coating antibody (see above) to each well of a 96-well plate.

2. Cover plate with pressure-sensitive film and incubate at room temperature for 1 h or at 4 °C overnight.

3. Aspirate each well and wash with at least 200 μl of wash buffer, repeat two times (three total washes) (*see* **Note 3**).

4. Add 100 μl of blocking solution to each well, cover plate with pressure-sensitive film and incubate at room temperature for 1 h (*see* **Note 4**).

5. Repeat washes as in **step 3**.

6. Add 100 μl each of hAAT standard (dilutions 1–9), samples, and positive controls to their respective wells (Fig. 2). Load one well with blocking buffer to serve as a "blank". Cover the plate

	1	2	3	4	5	6	7	8	9	10	11	12
A	Std 100ng/ml	50	25	12.5	6.25	3.125	1.5625	0.78125	0.390625	Blank	Pos 2X10E5	
B	Std 100ng/ml	50	25	12.5	6.25	3.125	1.5625	0.78125	0.390625	Blank	Pos 2X10E5	
C												
D												
E												
F												
G												
H												

Fig. 2 ELISA plate layout

with pressure-sensitive film and incubate at room temperature for 1 h (*see* **Note 5**).

7. Repeat washes as in **step 3**.

8. Add 100 μl of diluted secondary antibody to each well, cover the plate with pressure-sensitive film and incubate at room temperature for 1 h. Wash the plate six times with 200 μl of washing buffer per well.

9. Add 100 μl of TMB peroxidase substrate 1:1 dilution solution (*see* above) to each well. Develop the plate in the dark at room temperature for 30 min (*see* **Notes 6** and **7**).

10. Stop reaction by adding 50 μl of stop solution to each well (*see* **Note 8**).

11. Measure absorbance on a plate reader at 450 nm.

12. Calculate the concentration of unknown samples. Alternatively, computer-based curve-fitting statistical software may also be employed to calculate the concentration of the samples [2].

    (a) Calculate the mean absorbance for each set of duplicate standards, controls, and samples. Subtract the mean of blank control standard absorbance from each.

    (b) Construct a standard curve by plotting the mean absorbance for each standard on the y-axis against the concentration on the x-axis and draw a best fit curve through the points on the graph.

    (c) To determine the concentration of the unknowns, find the unknowns' mean absorbance value on the y-axis and draw a horizontal line to the standard curve. At the point of intersection, draw a vertical line to the x-axis and read the concentration. If samples have been diluted, the concentration read from the standard curve must be multiplied by the dilution factor.

4   Notes

1. It is very important to vortex all standard vials well between each dilution step and to vortex again as each standard is loaded to the well. Be cautious not to contaminate any sample wells with the standard samples while loading the plate.

2. Keep the samples on ice to thaw.

3. Do not let the plate dry out after coating antibody added. The plate can be blocked for a longer time until the standard and sample preparation is finished. Always leave the washing buffer in the well until the solution for the next step is ready.

4. Incubation at room temperature is about 22–25 °C.

5. Save at least one well for a blank control (load with blocking buffer). This will be used to determine the baseline background of the absorbance.

6. Bring the TMB peroxidase substrate to room temperature before use; this can be done by setting Solutions A and B on the bench an hour before it is needed (when the secondary antibody is added). Solution A and B must be protected from light.

7. Do not cover the plate with pressure-sensitive film on TMB peroxidase substrate development step. The solution must be kept in the dark for this step; we often accomplish this by placing the plate on a paper towel in a drawer during the incubation.

8. Read the plate within 30 min of the stop solution being added.

References

1. Song S, Scott-Jorgensen M, Wang J, Poirier A, Crawford J, Campbell-Thompson M, Flotte TR (2002) Intramuscular administration of recombinant adeno-associated virus 2 alpha-1 antitrypsin (rAAV-SERPINA1) vectors in a nonhuman primate model: safety and immunologic aspects. Mol Ther 6(3):329–335

2. Sono Biological Inc (online) ELISA protocol-calculation of results. ELISA encyclopedia http://www.elisa-antibody.com/general-elisa-protocol

Chapter 21

Quantification of Murine AAT by Direct ELISA

Andrew Cox and Christian Mueller

Abstract

This methods chapter elaborates on how a direct enzyme-linked immunosorbent assay (ELISA) is used to specifically detect and quantify murine alpha-1 antitrypsin (AAT). As a direct ELISA, it lacks some sensitivity as compared to the "sandwich" ELISA method; however, it does reliably differentiate between samples with varying amounts of the mouse AAT protein. This protocol relies on the principle of adsorption to coat each well with sera proteins, whereas detection occurs specifically using a two-step antibody combination. This procedure effectively identifies and quantifies murine AAT from a wide variety of samples including mouse serum, cell culture medium, and cell or tissue lysate.

Key words Alpha-1 antitrypsin, AAT, Murine AAT, Enzyme-linked immunosorbent assay, Direct ELISA

1 Introduction

The enzyme-linked immunosorbent assay (ELISA) is a quantitative protein plate-based assay analogous to immunodetection of proteins using a membrane [1]. More specifically, the ELISA quantifies a protein of interest that is in solution [2]. The direct ELISA, as compared to the "sandwich" ELISA, relies on the immobilization of a protein directly to the polystyrene well rather than coating with an initial capture antibody [3]. In this regard, the direct ELISA can be less sensitive as all sample components can bind with the well rather than just the protein of interest that will bind to the capture antibody [2]. In this protocol, two antibodies are used. The primary binds a specific epitope of the protein of interest, while the secondary (conjugated with a reporter) detects the primary [1]. After both interactions occur, the collective signal strength of the reporter is detected and quantified. Extrapolation from a standard curve and accounting for the sample dilution renders a quantifiable protein reading from the sample of interest.

This chapter outlines how to perform the murine AAT direct ELISA. As such, this protocol allowed for the identification and

Florie Borel and Christian Mueller (eds.), *Alpha-1 Antitrypsin Deficiency: Methods and Protocols*, Methods in Molecular Biology, vol. 1639, DOI 10.1007/978-1-4939-7163-3_21, © Springer Science+Business Media LLC 2017

differentiation of homozygous, heterozygous, and knockout mice that were bred in an AAT knockout study. Alpha-1 antitrypsin, as a secreted serpin protein, is naturally prevalent in sera samples, making it convenient to identify and quantify using the ELISA method. It is also possible to perform the same ELISA using cell culture media and tissue or cell lysates; however, expect to troubleshoot different sample dilutions [2].

2   Materials

Prepare and store all reagents at room temperature (unless otherwise indicated). Materials for waste disposal are to adhere to the institutional waste disposal regulations. All solutions are to be prepared using ultrapure water.

1. Clear flat-bottom 96-well microtiter plate.

2. Prepare 0.2 M sodium carbonate–bicarbonate coating buffer and adjust pH to 9.4 with NaOH. Autoclave and store at room temperature.

3. Prepare mouse AAT reference standard (*see* **Note 1**). Dilute mouse alpha-1 antitrypsin to 1000 ng/mL (Cat# RS-90A1T, Immunology Consultants Laboratory Inc., Portland, OR) in coating buffer. Store stock in 10 μL aliquots at −20 °C. Serially dilute to 500 ng/mL 1:2 (include 250 ng/mL, 125 ng/mL, 62.5 ng/mL, 31.25 ng/mL, 15.63 ng/mL, 7.81 ng/mL, and 3.9 ng/mL and 0 ng/mL).

4. Microplate sealing films.

5. Prepare 1× PBS-0.05% Tween 20 washing solution.

6. Prepare 2% skim milk blocking solution in 1× PBS–0.05% Tween.

7. Prepare unknown samples and controls (*see* **Notes 1** and **2**). Negative control as human serum (pooled normal human serum, Innovative Research and Technology, Cat# 1PLA-1). Make 50 μL aliquots and store at −80 °C. Positive control: serum from C57BL/6 mouse.

8. Prepare 1:2000 primary antibody (goat anti-mouse Alpha-1 Antitrypsin, Cat# GA1T-90A-Z, Immunology Consultants Laboratory Inc) in 2% skim milk blocking solution via 2-step dilution. Store stock solution at 4 °C.

9. Prepare 1:10,000 secondary detection antibody (HRP-conjugated rabbit anti-goat IgG h+l). Store stock at 4 °C.

10. Prepare freshly the TMB peroxidase substrate (TMB microwell peroxidase substrate system, Cat# 50-76-03, KPL Scientific, Gaithersburg, MD): mix equal volumes (5.5 mL each) of TMB

Peroxidase Substrate and Peroxidase Substrate Solution B. Store stocks at 4 °C.

11. Prepare 10% stop solution: 22.2 mL of 95–98% sulfuric acid, 377.8 mL of water.

12. Multichannel pipette.

13. Laboratory vacuum suction system.

14. Dilution plates.

15. Plate washer (optional), such as MultiWasher III (Cat# 8441-07, TriContinent, Grass Valley, CA).

16. Microplate reader with 450 nm filter, such as VersaMax (Molecular Devices, Sunnyvale, CA).

17. Software (optional), such as Softmax Pro (Molecular Devices).

3   Methods

1. Coat plate with 100 μL each of appropriately diluted standards (Table 1), blank (coating buffer), unknown samples, and controls (Table 2) to assigned wells (*see* **Notes 1** and **2**).

2. Cover microplate with film and incubate for 1 h at 37 °C (*see* **Note 3**).

Table 1
Murine AAT standards are prepared according to chart above. 1000 ng/mL starting solution is prepared using a two-step initial dilution

Step	Final AAT concentration (ng/mL)	AAT standard	Sample diluent (coating buffer) (μL)
2 step: 0	(a) 10,000 (b) 1000	1 μL 2.44 mg/mL stock 50 μL from step 0a	243 450
1	500	250 μL from step 0b	250
2	250	250 μL from step 1	250
3	125	250 μL from step 2	250
4	62.5	250 μL from step 3	250
5	31.25	250 μL from step 4	250
6	15.625	250 μL from step 5	250
7	7.8125	250 μL from step 6	250
8	3.90625	250 μL from step 7	250

Table 2
+/− controls and unknown sample dilution are prepared using this two-step 1:2500 dilution scale. Controls and unknown samples to fall within the concentration range of standards

Step	Final dilution	Sample	Sample diluent (coating buffer) (μL)
1	1:100	2μL of stock	198
2	1:2500	10μL from step 1	240

3. Wash plate 4× with 300 μL PBS–Tween 20 washing solution.

4. Aspirate remaining liquid in wells.

5. Add 100 μL of 2% skim milk 1×PBS–Tween 20 blocking solution to wells.

6. Cover with film and incubate for 1 h at 37 °C (*see* **Note 3**).

7. Wash plate 4× with 300 μL PBS–Tween 20 washing solution.

8. Aspirate remaining liquid in wells.

9. Add 100 μL of 1:2000 primary antibody.

10. Cover with film and incubate at 37 °C.

11. Wash plate 4× with 300 μL of PBS–Tween 20 washing solution.

12. Aspirate remaining liquid in wells.

13. Add 100μL of 1:10,000 secondary HRP conjugated detection antibody.

14. Cover with film and incubate at 37 °C.

15. Wash plate 4× with 300 μL PBS–Tween 20 washing solution.

16. Aspirate remaining liquid in wells.

17. Add 100 μL of mixed 1:1 TMB peroxidase substrate.

18. Incubate for 10–15 min at room temperature.

19. Add 50 μL of 10% stop solution.

20. Read microplate at OD of 450 nm.

21. Fit standard curve as trend line (4-parameter or linear regression) and calculate R^2 (*see* **Note 4**).

22. Use trend line fit curve to calculate the concentration of unknown samples (Figs. 1 and 2)

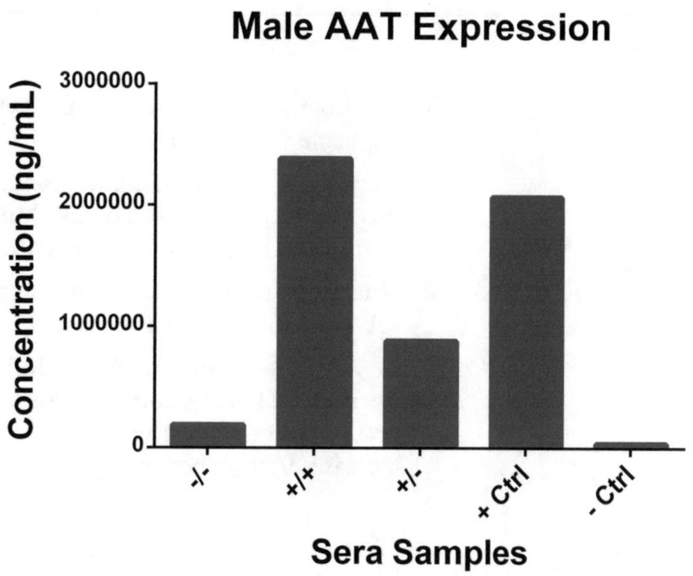

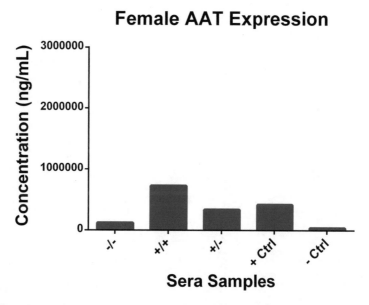

Fig. 1 Demonstrates an example of varying AAT levels in sera samples from male homozygous, heterozygous, and knockout mice. Positive Control: serum from male C57BL6 mouse, negative control: human serum. Data extrapolated from semilog linear standard

Fig. 2 Demonstrates an example of varying AAT levels in sera samples from female homozygous, heterozygous, and knockout mice. Positive Control: serum from female C57BL6 mouse, negative control: human serum. Data extrapolated from semilog linear standard

4 Notes

1. It is highly desirable to run standards and unknown samples in duplicate or triplicate.

2. Depending on the samples' concentration of AAT, dilutions may be necessary. Use coating buffer as diluting agent. Dilution has to be determined experimentally.

3. One hour-room temperature steps in this protocol can be substituted by incubating overnight at 4 °C.

4. R^2 with a 4-parameter fit should be ≥ 0.99. If using linear fit, use semilog scaling option, recommend $R^2 \geq 0.97$. Only interpret points falling within range of standard. Expand standard or increase dilution accordingly.

References

1. Qiagen Bench Guide (2001) Protein Assay, 83–87

2. Ausubel F et al (2003) Current protocols in molecular biology. John Wiley & Sons, New York. (pages 1661, Section 11.2.1-22)

3. KPL Technical Guide for ELISA (2013) Assay Formats, 5–7

Chapter 22

Quantification of Z-AAT by a Z-Specific "Sandwich" ELISA

Florie Borel, Qiushi Tang, and Christian Mueller

Abstract

This protocol describes an enzyme-linked immunosorbent assay (ELISA) to specifically detect Z-alpha-1 antitrypsin (AAT), the most common protein variant associated with alpha-1 antitrypsin deficiency. This "sandwich" ELISA relies on an anti-Z-AAT specific capture antibody and a HRP-conjugated anti-AAT detection antibody. This method would be of interest to identify and quantify Z-AAT in a variety of samples such as cell culture medium, cell or tissue lysate, animal or patient serum. Because this method is specific and sensitive, it would be particularly valuable for detection of Z-AAT in the presence of background M-AAT, for instance when quantifying silencing of Z-AAT in patients undergoing M-AAT augmentation therapy.

Key words Alpha-1 antitrypsin, AAT, Z-AAT, Enzyme-linked immunosorbent assay, Sandwich ELISA, UMMS-PiZ-mAb

1 Introduction

Enzyme-linked immunosorbent assay (ELISA) is a plate-based assay to quantify a protein of interest. While various formats of ELISA exist, the so-called "sandwich" assay relies on a capture antibody which will bind the protein to the plate, after which any unbound antigen is removed by washes, and a detection antibody which is conjugated to an enzyme, leading to signal readout.

In this chapter we describe how to perform a "sandwich" ELISA to detect Z-AAT, the most common protein variant associated with leading to alpha-1 antitrypsin deficiency. Alpha-1 antitrypsin is a secreted protein, its detection in animal or patient serum is therefore a convenient readout for identification or quantification of the protein, but cell culture medium or cell or tissue lysates can also be used. In this protocol, we use a Z-AA-specific IgG$_{2a}$ Kappa antibody that recognizes a linear Z epitope [1], which has been shown to detect both Z-AAT-elastase complexes and polymerized Z-AAT [2]. The specificity of this assay is demonstrated in Fig. 1, and its sensitivity in Fig. 2 (dynamic range: 0.150–1.250).

Florie Borel and Christian Mueller (eds.), *Alpha-1 Antitrypsin Deficiency: Methods and Protocols*, Methods in Molecular Biology, vol. 1639, DOI 10.1007/978-1-4939-7163-3_22, © Springer Science+Business Media LLC 2017

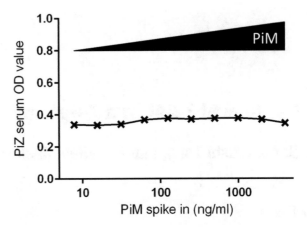

Fig. 1 Specificity of the assay. The assay was performed as described in the protocol, using a fixed amount of Z-AAT and increasing amounts of M-AAT. Variation in OD value is minimal when an increasing amount of PiM protein is spiked in, demonstrating the sensitivity of this method

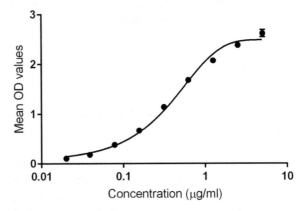

Fig. 2 Sensitivity of the assay. The assay was performed as described in the protocol, with increasing amounts of Z-AAT in the protein standard. Values for the standard are: 0.020, 0.039, 0.079, 0.156, 0.313, 0.625, 1.250, 2.5, 5. As shown here, the linear range of the standard curve is between 0.150 and 1.250, which represents the dynamic range for this assay. The equation for the 4PL nonlinear regression is $Y = (A - D)/1 + (X/C)^B) + D$, where A is the minimum asymptote, B the hill slope, C the inflexion point, and D the maximum asymptote

2   Materials

Prepare and store all reagents at room temperature (unless indicated otherwise). Dispose of waste materials according to institutional waste disposal regulations. Prepare all solutions using ultrapure water.

1. Clear flat-bottom nonsterile 96-well plates.

2. Voller's buffer: 2.76 g sodium carbonate (Na_2CO_3, 106.0 g/mol), sodium bicarbonate 1.92 g ($NaHCO_3$,

84.01 g/mol), 0.2 g sodium azide (NaN$_3$, 65.01 g/mol). Add water to 1 l. Adjust pH to 9.6 with NaOH. Autoclave and store at room temperature.

3. Prepare 1:100 dilution of capture antibody: 105 μl of mouse monoclonal anti-human Z-alpha-1 antitrypsin clone UMMS-PiZ-mAb (produced by our lab, for orders *see* www.umassmed.edu/muellerlab/reagent-request), 10.5 ml of Voller's buffer.

4. Falcon microplate sealing film.

5. PBS–Tween: 500 μl of Tween 20, 1 l of 1× PBS.

6. 5% BSA in PBS.

7. Prepare Z-AAT standard (*see* **Note 1**). Dilute PiZ mouse serum to 500 μg/ml (Lot#PIZ-001, UMass Medical School) in 5% BSA. Store 20 μl aliquots at −80 °C. Dilute the 500 μg/ml aliquot 1:25 to get 20 μg/ml. Subsequently, prepare 1:2 serial dilutions.

8. Prepare unknown samples (*see* **Notes 1** and **2**).

9. Prepare 1:10 detection antibody (goat polyclonal anti-human alpha-1 antitrypsin, HRP-conjugated, Cat#ab191350, Abcam, Cambridge, UK) in 5% BSA. Store 23 μl aliquots at −80 °C. Prepare 1:5000 detection antibody: 22 μl of 1:10 dilution of detection antibody, 11 ml of 5% BSA.

10. TMB peroxidase substrate: TMB microwell peroxidase substrate system (Cat#50-76-00, SeraCare Life Sciences, Milford, MA). Prepare fresh by mixing 6 ml of reagent A with 6 ml of reagent B.

11. Stop solution: 22.2 ml of 95–98% sulfuric acid, 377.8 ml of water. Store at room temperature.

12. 12-channel pipette.

13. Plate washer (optional).

14. Microplate reader with 450 nm filter, such as VersaMax (Molecular Devices, Sunnyvale, CA).

15. Software (optional), such as SoftMax Pro (Molecular Devices).

3 Methods

1. Coat plates with 100 μl of 1:100 capture antibody.

2. Cover with film and incubate O/N at 4 °C.

3. Wash plate 3× with 200 μl PBS–Tween.

4. Add standard (*see* **Note 1**).

5. Add unknown samples (*see* **Notes 1** and **2**).

6. Cover with film and incubate for 1 h at 37 °C.

7. Wash plate 3× with 200 μl PBS–Tween.

8. Add 100 μl of 1:5000 detection antibody.

9. Cover with film and incubate for 1 h at 37 °C.

10. Wash plate 6× with 200 μl PBS–Tween and make sure to remove any remaining PBS–Tween after the last wash.

11. Add 100 μl TMB peroxidase substrate.

12. Incubate for approximately 30 min at room temperature.

13. Add 50 μl stop solution.

14. Read OD450.

15. Fit standard curve data as a trend line (four-parameter logistic nonlinear regression) and calculate R^2 (*see* **Note 3**).

16. Calculate the concentration of unknown samples.

4   Notes

1. It is recommended to run the standard and the unknown samples in triplicate. Thaw samples on ice. Make sure all samples and dilution steps are vortexed very well. Leave one well with coating buffer only to use as a blank control.

2. Depending on the concentration of the samples, it may be necessary to dilute them in 5% BSA. This should be determined experimentally. The unknown concentrations have to fall within the dynamic range of the assay (*see* Fig. 2).

3. R^2 value should be ≥ 0.95.

Acknowledgments

Development of the PiZ-specific antibody was funded by the Alpha-1 Foundation.

References

1. Wallmark A, Alm R, Eriksson S (1984) Monoclonal antibody specific for the mutant PiZ alpha 1-antitrypsin and its application in an ELISA procedure for identification of PiZ gene carriers. Proc Natl Acad Sci U S A 81 (18):5690–5693

2. Janciauskiene S, Dominaitiene R, Sternby NH, Piitulainen E, Eriksson S (2002) Detection of circulating and endothelial cell polymers of Z and wild type alpha 1-antitrypsin by a monoclonal antibody. J Biol Chem 277(29):26540–26546. doi:10.1074/jbc.M203832200

Semiquantitation of Monomer and Polymer Alpha-1 Antitrypsin by Centrifugal Separation and Assay by Western Blot of Soluble and Insoluble Components

Keith S. Blomenkamp and Jeffrey H. Teckman

Abstract

Alpha-1 antitrypsin (a1AT) deficiency, in its classical form, is an autosomal recessive disease associated with an increased risk of liver disease in adults and children, and with lung disease in adults. The vast majority of liver disease is associated with homozygosity for the Z mutant allele, also called PiZZ. This homozygous allele synthesizes large quantities of a1AT mutant Z protein in the liver, but the mutant protein also folds improperly during biogenesis. As a result, approximately 85% of the molecules are retained within the hepatocytes instead of being appropriately secreted. The resulting low, or "deficient," serum level leaves the lungs vulnerable to inflammatory injury from uninhibited neutrophil proteases. Most of the mutant Z protein retained within hepatocytes is directed into intracellular proteolysis pathways, but some molecules remain in the endoplasmic reticulum for long periods of time and others adopt an unusual aggregated or "polymerized" conformation. It is thought that these intracellular polymers trigger a cascade of intracellular injury which can lead to end organ liver injury including chronic hepatitis, cirrhosis, and hepatocellular carcinoma. It is widely accepted that the disease causing factor in mutant Z-alpha-1 antitrypsin deficiency (AATD-Z) is the toxic build-up of the mutant Z protein. Since misfolding of some but not all of the Z protein during its maturation leads to homopolymerization, an assay to assess the amount of normally folded ATZ and accumulated polymeric ATZ would be very useful. Here we describe a method to semiquantitatively assess these two fractions in a tissue or cell culture source.

Key words Alpha-1 antitrypsin, AAT, ATZ, Z mutant, Liver disease, Liver cirrhosis, Storage disease, Aggregation, Polymerization, Soluble/insoluble protein separation, Monomer–polymer

1 Introduction

These methods were adapted from Dul et al. [1] and Lin et al. [2]. This assay makes use of a low detergent lysate buffer and homogenization technique that disrupts the cellular membranes without dissociating the ATZ polymers. A small aliquot of this lysate can be subjected to centrifugation, leaving a pellet and supernatant that can be separated without detectable cross-contamination. The soluble fraction is thought to consist of the majority, if not all, of the

Florie Borel and Christian Mueller (eds.), *Alpha-1 Antitrypsin Deficiency: Methods and Protocols*, Methods in Molecular Biology, vol. 1639, DOI 10.1007/978-1-4939-7163-3_23, © Springer Science+Business Media LLC 2017

monomeric ATZ protein. Conversely, the insoluble fraction is considered to be almost all of the polymeric and aggregated ATZ protein. These fractions can then be used in a variety of analyses. The one described here is a denaturing sodium dodecyl sulfate–polyacrylamide gel electrophoresis (SDS-PAGE) and immunoblot Western blot that reduces essentially all of the ATZ to its monomeric form, thereby enabling a direct quantitative comparison.

2 Materials

1. MP lysate buffer: 50 mM Tris–HCl, pH 8.0, 150 mM NaCl, 5 mM KCl, 5 mM $MgCl_2$, 0.5% Triton X-100, 1:100 Sigma Protease Inhibitor Cocktail, 500 mM EDTA, 500 mM PMSF, 500 mM 1,10-phenanthroline, 500 mM iodoacetamide (*see* **Note 1**).

2. Resolving gel: 10% acrylamide, 0.375 M Tris–HCl, pH 8.8, 0.1% SDS.

3. Stacking gel: 4% acrylamide, 0.125 M Tris–HCl, pH 6.8, 0.1% SDS (*see* **Note 2**).

4. 5× sample buffer: 0.5 M Tris–HCl, pH 6.8, 5% SDS, 10% β-mercaptoethanol (BME), 50% glycerol, pinch of bromophenol blue.

5. Running buffer: 25 mM Tris, 192 mM glycine, pH 8.3, 0.1% SDS.

6. Transfer buffer: 25 mM Tris, 192 mM glycine, pH 8.3, 20% methanol.

7. PBS-T wash buffer: 1× phosphate buffered saline (PBS), 0.1% Tween 20 (PBS-T).

8. Blocking buffer: 1× PBS, 0.1% Tween 20, 5% nonfat milk powder.

9. Cells or tissue source: Fresh or frozen cells or frozen tissue can be used (*see* **Note 3**).

10. Homogenization tools: Tweezers, razor blade, weigh paper, two ice buckets, ice, dry ice, glass Dounce homogenizer with a tight pestle, 1 ml syringe with a 28-gauge needle, 10–1000 μl single channel pipettor, tips, analytical scale.

11. Separation materials: Refrigerated centrifuge, 1–20 μl single channel pipettor, gel loading tips, vortex.

12. Boiling materials: Sample buffer, hot plate or flame, float, vessel, tabletop centrifuge Gel: Molecular weight marker, 10% gel with 4% stacking gel, gel loading pipette tips, gel resolving apparatus.

13. Transfer materials: PVDF or nitrocellulose membrane, transfer apparatus, cold room.

14. Probing: Rocker/shaker, PBS-T, membrane boat, blocking buffer, anti-alpha-1 as primary antibody, wash buffer, appropriate secondary HRP conjugated antibody.

15. Detection materials: ECL Plus or similar, Laser scanner or X-ray film, plastic wrap, tweezers.

16. Analysis tools: Computer, and ImageJ or similar image analysis software.

3   Methods

3.1   A1AT Monomer–Polymer Separation of PiZ Mouse Liver

1. All steps should be performed on ice until samples are denatured in the boiling step.

2. Obtain a whole liver sample. Weigh out approximately 10 mg of liver, 5 mg of tissue per 1 ml of MP Buffer and keep it chilled. (*see* **Note 1, 3, 8**, and **14**)

3. Add liver tissue at a 1:200 w/v ratio to the MP Homogenate Buffer containing protease inhibitors.

4. Homogenize with chilled, glass Dounce homogenizer with a tight pestle for 40 repetitions. This step is optional if a cell culture source is used. (*see* **Note 4**)

5. Vortex vigorously and aliquot 0.5–1 ml (0.5 ml works well) into a fresh labeled tube for next step. Keep on ice. The remainder can be stored or discarded as needed.

6. Using a 1 ml syringe with a 28-gauge needle, pass the homogenate through (in and out) of the needle ten times while keeping tube cold (*see* **Note 4** and **9**).

7. Perform a standard total protein determination on syringe disrupted sample(s) (*see* **Note 10**).

8. Calculate total micrograms of protein per microliter. If you adhered to the 5 mg liver to 1 ml buffer ratio, you should expect a total protein concentration of approximately 0.8–1.2 µg/µl.

9. Pipette a calculated volume of protein that is equal to 5 µg of total protein into a standard 1.5 ml microcentrifuge tube. Do not use coated or super low adhesion tubes. Label this tube "Polymer" or "Insoluble" (*see* **Note 5** and **11**).

10. With a balanced angle rotor, spin the quantified amount at $15,000 \times g$ for at least 30 min at 4 °C. The pellet is usually invisible or difficult to see. Therefore, it is important to position the tubes in a consistent manner so that the pellet location can be estimated and avoided during aspiration of supernatant.

11. Set the pipette to 2.5 µl higher than the volume of sample that is spinning. When spinning is complete, immediately pull off all of the supernatant by using a long stemmed gel loader pipette

tip, being extremely careful not to disturb a sometimes invisible pellet and return pellet/insoluble tube to the ice. The tip should pull up a small amount air after all of the supernatant has been aspirated (*see* **Note 6, 12**, and **13**).

12. Turn down the volume on the dial until the air is dispensed.

13. Measure and record the volume.

14. Deposit the supernatant into another tube labeled "Monomer" or "Soluble." Set aside in ice (*see* **Note 6**).

15. Add at least 5 μl of 5× Sample Buffer to each monomer tube and at least 10 μl of 5× Sample Buffer to each polymer tube.

16. Vortex the pellet vigorously for 20–30 s to ensure that the pellet is dislodged and resuspended.

17. Incubate the tubes for 5 min in boiling water.

18. Vortex the tubes once more vigorously for 10 s.

19. Let the tubes cool and then quickly spin the samples in order to collect them at the bottom.

20. Load all of the samples onto the gel. This is semiquantitative and, therefore, it is very important that all of the sample is loaded and none is lost (residue is negligible) (*see* **Note 7**).

21. The gel should be run at 120 V with an 85 mA current limit until the dye front reaches the bottom or the 50 kDa molecular weight marker is at the vertical midline. Wet transfer is run at 100 V at a 250 mA limit for 1.5 h.

3.2 Western Blot Probe

1. Block the blot in 5% nonfat milk that contains 1× PBS 0.1% Tween 20 (PBS-T) at room temp on the rocker for 1 h or overnight at 4 °C on the rocker.

2. Add the primary antibody, DiaSorin Goat Anti-human Alpha-1 antitrypsin (item discontinued, alternative may be available) at a concentration of 1:50,000 in PBS-T to the blot and place on the rocker for 23 min at room temperature. Other antibodies specific to AAT can also be used.

3. Rinse the blot twice and wash it three times in PBS-T for 10 min each time at room temperature on the rocker.

4. Add the secondary antibody, DAKO Rabbit anti-Goat IgG HRP conjugated at a concentration of 1:100,000 in PBS-T to the blot and place on the rocker for 25 min at room temperature. Other antibodies that are compatible with the primary antibody can also be used.

5. Rinse the blot twice and wash it three times in PBS-T for 10 min each time at room temperature on the rocker.

6. Incubate each blot with 2 ml of the ECLplus detection reagent for 5 min, rinse with ddH$_2$O, and scan the blot with the GE Typhoon 9410 laser scanner using appropriate excitation and emission settings. Alternatively, chemiluminescent reagents and film can be used (Fig. 1). (*see* **Note 15, 16**, and **17**).

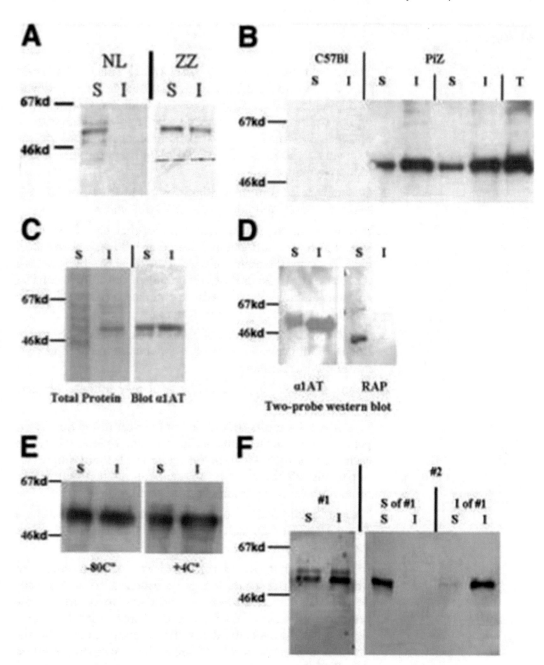

Fig. 1 An et al. [3]. Analysis of α1ATZ insoluble polymers, SDS-PAGE and immunoblot with alpha polyclonal anti-α1AT antibody of S, I, and total fractions of wild-type, MM human liver (NL), homozygous ZZ human liver, PiZ transgenic mouse liver, and wild-type mouse liver (C57BL). (**a**) Normal versus ZZ. (**b**) C57Bl versus PiZ mice. (**c**) Whole protein stain versus immunoblot of ZZ liver. (**d**) PiZ mouse liver sequential immunoblot first with anti-α1AT antibody and then with anti-RAP antibody. (**e**) Aliquots of the same ZZ liver homogenate kept at −80 °C for 1 week versus 4 °C for 1 week and blotted for α1AT. (**f**) ZZ liver subjected to the isolation procedure once (#1), and then aliquots of the S and I fractions subjected to a second round of the procedure (#2). NL, normal; S, soluble; I, insoluble; T, total; α1AT, alpha-1 antitrypsin; RAP, receptor-associated protein

4   Notes

1. The reagents 50 mM Tris–HCl, pH 8.0, 150 mM NaCl, 5 mM KCl, 5 mM $MgCl_2$, 0.5% Triton X-100 should be mixed with freshly added 1:100 Sigma Protease Inhibitor Cocktail or a similar solution. This mixture can be supplemented with an optional 500 mM (5 mM working concentration) of the following: EDTA, PMSF, 1,10-phenanthroline, and iodoacetamide.

2. Generally, a 1.5 mm thick minigel is used with 10 or 15 wells that are capable of holding at least 20 μl volume.

3. If using a stored, frozen liver, samples that have been flash-frozen in liquid nitrogen and stored at −80 °C are the most ideal. A practical amount of starting material would be at least 5 mg of tissue or 25 cm^2 of 90% confluent cell culture. Smaller amounts are not advised, because sampling error or minimum level of detection may be encountered.

4. Try to avoid creating a lot of bubbles as this will make aspiration more difficult. Some bubbles will be unavoidable and will not hinder performance.

5. Volumes above 10 μl or below 4 μl may inhibit complete detectable separation.

6. Unseparated samples should not sit after centrifugation as the pellet may be released into the supernatant. Spinning between each tube's sample-separation is necessary. It is also important not to disturb the rotor or adjacent tubes when removing a tube. Re-balance rotor after each removal.

7. This procedure is done on a Bio-Rad minigel system using a 10- or 15-well comb with 4% PAGE-SDS stacking gel and 10% PAGE-SDS running gel.

8. In the case of an in vivo source, in order to avoid sampling error when making the lysate, it may be beneficial to have already done a histological analysis on the tissue to determine the presence and scale of ATZ globule heterogeneity. If there is significant heterogeneity, it may be necessary to increase the samples size or number of areas. The sampling method described above will be appropriate for all but the most extreme cases.

9. During the syringe disruption, small particulates, probably connective tissue, may be encountered. If they clog the needle, try purging the syringe, starting over, and attempting to resuspend again. However, if there is a piece that will not pass and it is preventing the procedure, it can be carefully removed from the needle tip and discarded altogether.

10. If the resultant protein concentrations for a given experimental set have a large range it may be necessary to normalize them before separation to ensure accurate results.

11. Avoid aspirating any visible lysate particulates when aliquoting for blot samples that will be centrifugally separated. Simply aspirate around them through visual observation. If it is not big enough to see, it will not be a problem.

12. When using the gel loading pipette to aspirate the supernatant from the spun tube, hold the tube vertically. This will maximize supernatant removal. The alternative is holding the tube at an angle with the pellet high on the side and the tip horizontal, although this could leave some supernatant behind. It also helps to hold the pipette tip vertically and go in slowly on the opposite side of where the pellet would reside, finally reaching the absolute bottom of the conical tube before aspirating slowly but all at once.

13. The separation step seems to have a relatively high level of difficulty for the beginner. Some practice and quality control of performance may be necessary before delving into the actual final assay design. Running some sample repetitions is a valuable way of highlighting separation success and consistency.

14. When using cell culture as the sample source, the protocol may need to be optimized. Each cell culture may have significantly different amounts of soluble and insoluble ATZ. Titration may be necessary. A good starting point for a cell culture sample is about 30–60 µg total protein if it is collected in the manner outlined above. The primary and secondary antibody dilutions and their incubation times may need to be optimized as well. Starting points for these would be: primary antibody 1:5000 for 1 h at room temperature or overnight at 4 °C; secondary antibody 1:50,000 for 1 h at room temperature.

15. A reference antibody probe should be used to determine consistent loading. However, most references are soluble only and will not be present in the insoluble lanes. The respective soluble component's loading reference can be used in the place of this absence.

16. The two possible forms of monomeric soluble AAT, or in this case ATZ, that will be observed on the blot will be the mature form weighing 55 kDa and/or the immature form weighing 52 kDa.

17. There are two potential ways to assess complete fraction separation. Depending on the level of signal, there is almost always some denature-resistant higher molecular weight AAT-reactive species in the insoluble fractions. This is normal. However, if you see this in its respective samples soluble fraction blot lane, it is likely that the insoluble pellet has been disturbed during

supernatant aspiration and contaminated the soluble fraction. Conversely, the mature form of ATZ, observed at 55 kDa on the blot, is not usually found in the insoluble fraction's blot lane. If it is found in this lane, it is likely that some of the supernatant has been left behind with the insoluble fraction. Only the 52 kDa form should be associated with SDS/BME-dissociated polymer found in the successfully separated insoluble fraction.

References

1. Dul JL, Davis DP, Williamson EK, Stevens FJ, Argon Y (2001) Hsp70 and antifibrillogenic peptides promote degradation and inhibit intracellular aggregation of amyloidogenic light chains. J Cell Biol 152(4):705–716

2. Lin L, Schmidt B, Teckman J, Perlmutter DH (2001) A naturally occurring nonpolymerogenic mutant of alpha 1-antitrypsin characterized by prolonged retention in the endoplasmic reticulum. J Biol Chem 276(36):33893–33898

3. An JK, Blomenkamp K, Lindblad D, Teckman JH (2005) Quantitative isolation of alpha1AT mutant Z protein polymers from human and mouse livers and the effect of heat. Hepatology 41(1):160–167

Chapter 24

Electrophoresis- and FRET-Based Measures of Serpin Polymerization

Sarah V. Faull, Anwen E. Brown, Imran Haq, and James A. Irving

Abstract

Many serpinopathies, including alpha-1 antitrypsin (A1AT) deficiency, are associated with the formation of unbranched polymer chains of mutant serpins. In vivo, this deficiency is the result of mutations that cause kinetic or thermodynamic destabilization of the molecule. However, polymerization can also be induced in vitro from mutant or wild-type serpins under destabilizing conditions. The characteristics of the resulting polymers are dependent upon induction conditions. Due to their relationship to disease, serpin polymers, mainly those formed from A1AT, have been widely studied. Here, we describe Förster resonance energy transfer (FRET) and gel-based approaches for their characterization.

Key words Polymerization, Protein aggregation, Kinetics, Monoclonal antibody, Immunoblots

1 Introduction

The serpinopathies are a group of hereditary diseases associated with pathological gain-of-function or loss-of-function phenotypes. The underlying bases of many of these conditions are the misfolding and accumulation of mutant serpins, including alpha-1 anti-trypsin (A1AT) [1]. In vivo, mutant A1AT forms linear, flexible unbranched chains of molecules with a "beads-on-a-string" morphology when visualized by electron microscopy (EM) [2, 3]. These are referred to as polymers. Their accumulation results in material deposits at the site of synthesis, observable in ex vivo samples as diastase resistant and Periodic Acid Schiff positive inclusions within hepatocytes [4]. The consequence of the deposits is physical distension of the endoplasmic reticulum, where the protein is synthesized as the first step in the secretory pathway, along with consequences for intraluminal mobility [5]. However, polymer formation does not trigger the unfolded protein response that is associated with an accumulation of misfolded proteins [5–7].

The structural configurations that are adopted by serpins can be classified into two distinct states: (a) the kinetically stable, but

Florie Borel and Christian Mueller (eds.), *Alpha-1 Antitrypsin Deficiency: Methods and Protocols*, Methods in Molecular Biology, vol. 1639, DOI 10.1007/978-1-4939-7163-3_24, © Springer Science+Business Media LLC 2017

thermodynamically unstable native conformation, with a 5-stranded central β-sheet A and an exposed reactive center loop; and (b) the thermodynamically hyperstable 6-stranded β-sheet A "reactive center loop incorporated" conformation [8, 9]. It is proposed that the subunits of the polymer conform to the latter configuration [2, 10, 11]. Accordingly, polymers exhibit extreme stability [12].

The key to understanding the mechanism of polymerization involves the development of methods that are capable of inducing it under controlled conditions in vitro. The utilization of such methods arises in part from a relative scarcity of explant liver material, and they also provide a means of manipulating polymer formation as a kinetic process. These methods have typically involved the use of conditions that destabilize the native state: (a) denaturant [2]; (b) heat [12]; or (c) acidic pH [13]. However, different conditions result in polymers that are distinct immunologically [14, 15] as well as in structural character [11, 13]; findings that suggest different polymer forms are accessible through in vitro methods. This has likely been a contributing factor in the emergence of several apparently inconsistent models of the pathological A1AT polymer [2, 3, 10, 11]. Consistent with the observation that peptides based on this region were found to antagonize the polymerization process [2], as well as the observation that polymers show comparable stability to the loop-incorporated form [16], these models all involve a reactive center loop embedded in a 6-stranded β-sheet A, although they differ in their representation of the associated intermolecular linkage.

Biochemically, the process of polymerization can be described by the following Scheme [17]:

$$M(\text{monomer}) \rightarrow M^*(\text{activated monomer})$$
$$M^* + M^* \rightarrow P(\text{polymer})$$

in which the "activated monomer" represents one or more monomeric, oligomerization-competent intermediates on the pathway [18, 19].

Given the choice of destabilizing conditions influencing the nature of the resultant polymers, a key question involves the degree to which observations made in vitro can be extended to the pathological context. This is exemplified by the 2C1 monoclonal antibody, which recognizes an epitope held in common between pathological material and heat-induced, but not denaturant-induced, in vitro polymers [14]. Furthermore, there is debate as to whether polymers arise in vivo from a near-native state or a folding intermediate. The latter possibility appears to be consistent with the observation that the denaturant-mediated unfolding pathway traverses an intermediate ensemble that has a tendency to polymerize [2, 16]. However, this intermediate shows a measurably lower degree of packing with respect to that which is imparted during heating [20, 21]. This is important because different studies have made use of these two techniques in order to draw conclusions about polymerization processes that may in fact be distinct.

An understanding of the process of polymerization is of direct relevance to the development of therapeutics [22] and studies aimed at characterizing or manipulating the cellular response to the polymer burden [5, 23]. In this chapter, we provide methods for assessing the progress of heat-induced A1AT polymerization.

2   Materials

High-purity water denotes >10 MΩ resistivity; ultrapure denotes >18 MΩ resistivity at 25 °C.

2.1 Preparation of Fluorescent A1AT

1. A spectrophotometer capable of measuring absorbance in the UV and visible range.

2. A UV transilluminator with camera assembly or a fluorescence gel scanner.

3. Disposable centrifugal concentrator units with a cutoff of 10 kDa or 30 kDa (the latter will provide faster concentrations).

4. Dialysis membrane with a cutoff between 10 kDa and 30 kDa.

5. Syringe or centrifugal filters (with 0.45 μm cutoff).

6. Phosphate-buffered saline with sodium azide (PBSZ): 8 g NaCl, 0.2 g KCl, 1.44 g Na_2HPO_4, 0.24 g KH_2PO_4, 0.2 g sodium azide. (Made to a volume of 0.8 l in ultrapure water, the pH adjusted as necessary to 7.4 with HCl, volume adjusted to 1 l with ultrapure water, and filtered through a 0.45 μm filter).

7. 1 M Dithiothreitol in ultrapure water. Divided into aliquots and frozen.

8. 1 M L-cysteine hydrochloride in ultrapure water.

9. Two thiol-reactive fluorescent labels, one that will be directly excited (the "donor") and one indirectly excited (the "acceptor"). For thiol selectivity and stability, maleimide-based conjugates are recommended (see **Note 1**). These are diluted to a concentration of 50 mM in DMSO and stored in small aliquots (enough for one estimated use) at −80 °C.

10. Human plasma A1AT (see **Note 2**), or human recombinant A1AT with either the wild-type cysteine at position 232 or another solvent-accessible cysteine introduced by mutagenesis. Protein should have been purified to homogeneity (see **Note 3**).

2.2 Sample Preparation Using Anion Exchange Chromatography

1. A chromatography system (pump, fraction collector, disposable tubes) (*see* **Note 4**).
2. A compatible 1 ml HiTrap Q column or an equivalent strong anion exchange resin.
3. A UV transilluminator with camera assembly or a fluorescence gel scanner.
4. Anion exchange buffer A: 20 mM Tris pH 8.0, 0.2 g sodium azide. (Made in 1 l ultrapure water and filtered through a 0.45 μm filter; *see* **Note 5.**)
5. Anion exchange buffer B: 20 mM Tris pH 8.0, 1 M NaCl, 0.2 g sodium azide. (Made in 1 l ultrapure water, and filtered through a 0.45 μm filter; *see* **Note 5.**)

2.3 Heat-Induced Polymerization Monitored by FRET

1. Real-time thermal cycler (*see* **Note 6**).
2. Centrifuge with plate holders.
3. Real-time compatible PCR plate and plate sealant.
4. Suitable sample buffer, at the desired pH and ionic strength, such as PBS.
5. Two derivatives of plasma-derived or recombinant A1AT: one labeled using a "donor" fluorophore and one labeled using an "acceptor" fluorophore, at a stock concentration of 1 mg ml^{-1} or greater in PBSZ.

2.4 End-Point Experiments of Polymerization

1. A thermal cycler capable of a generating a temperature gradient across the plate (fluorescence detection is not necessary).
2. A UV transilluminator or fluorescence gel scanner if fluorescence detection is used (optional).
3. Centrifuge with plate holders.
4. PCR plate or tubes with sealant or lids.
5. Suitable sample buffer, at the desired pH and ionic strength, such as PBSZ (as described above).
6. 5× native loading buffer: 5 ml glycerol, 5 ml ultrapure water, 10 mg bromophenol blue.
7. 3–12% bis-Tris NativePAGE™ gels with running buffer or an equivalent native polyacrylamide gel system using Tris/Glycine buffers.
8. Coomassie or SYPRO Orange™ gel stain to visualize bands.
9. Plasma-derived or recombinant A1AT, unlabeled or labeled at the single cysteine with a fluorescein derivative (*see* **Note 7**).

2.5 Polyacrylamide-Agarose Hybrid "Slab" Gels

1. Horizontal slab-gel DNA electrophoresis system, including the tank, combs, and gel casting box.
2. Fluorescence gel scanner.

3. 30% (37.5:1) acrylamide/bis-acrylamide solution (*see* **Note 8**).

4. 10× cathode buffer: 64 g l^{-1} Tris base, 51 g l^{-1} glycine, high purity water.

5. Tetramethylethylenediamine (TEMED).

6. Agarose powder suitable for DNA electrophoresis.

7. 10% ammonium persulfate solution, freshly prepared.

8. 5× native loading buffer: 5 ml glycerol, 5 ml ultrapure water, 10 mg bromophenol blue.

9. SYPRO Orange™ gel stain.

10. 7.5% glacial acetic acid, 10% SDS, in ultrapure water.

11. 7.5% glacial acetic acid in ultrapure water.

3   Methods

3.1 Preparation of Fluorescent A1AT

Sequestration of individual molecules into a growing polymer chain brings them into close proximity. Therefore, Förster resonance energy transfer (FRET) can be used to monitor the progress of the reaction in real time. This requires two preparations of A1AT with different fluorescence properties to be prepared, using "donor" and "acceptor" fluorophore conjugates. Follow the protocol detailed here for each fluorophore (*see* **Note 1**).

1. Recombinant human A1AT (typically expressed in *E. coli*), or purified human plasma A1AT, harbors an endogenous cysteine at position 232. For optimal thiol conjugation efficiency, reduce the protein prior to the introduction of a fluorescent conjugate for 10–30 min at room temperature with 10 mM dithiothreitol (*see* **Note 2**).

2. Remove excess reducing agent by using a subsequent ≥1000-fold buffer exchange through dialysis. A maleimide conjugation reaction is most efficient at a neutral pH, thus phosphate buffered saline azide (PBSZ) is a suitable buffer to exchange the sample into. As an alternative to dialysis, anion exchange chromatography as described below can be performed (*see* **Note 3**). If the sample is purified in this way, dilute 1:3 v/v into PBSZ.

3. The concentration of a pure A1AT preparation can be determined by measuring the absorbance of a sample at 280 nm in a spectrophotometer, either cuvette-based or with a device such as a NanoDrop™. The absorbance of a 1 mg ml^{-1} solution of plasma-derived A1AT in a 1 cm pathlength is 0.52, while that of the recombinant protein is 0.58. Adjust the concentration to around 0.5–2 mg ml^{-1} in PBSZ by dilution or concentration using centrifugal concentrator units with a 10 kDa or 30 kDa cutoff.

4. Divide the protein that will be labeled into two tubes. While protecting the reaction from light (*see* **Note 9**), add a 20-fold molar excess of each fluorescent conjugate to the protein concentration. Incubate the samples at 25 °C for 18 h, in the dark. The incubation can be continued for an additional 24 h with no adverse effect.

5. Quench the labeling reaction through the addition of sufficient L-cysteine from a 1 M stock solution until the concentration reaches 10 mM. Use a 0.45 μm filter, either syringe-based or a centrifugal unit, to remove any precipitate.

6. Exchange the sample ≥1:200 into PBSZ by using a centrifugal concentrator unit, preferably with a 30 kDa cutoff. If available, anion exchange chromatography as described below also provides an effective means to remove most of the remaining unconjugated fluorophore.

7. The presence of the unconjugated label can be determined by resolving the resulting samples through SDS-PAGE, and visualizing them with a UV transilluminator or fluorescence gel scanner (such as a Typhoon Trio, GE Healthcare) without staining. The free fluorophore will migrate close to the dye front. Determine the fluorescence intensity of the band near the dye front as a proportion of the labeled protein, using densitometry software such as Image J.

8. To minimize the background in FRET experiments, the unconjugated label would ideally be barely discernible, but should comprise less than 10% of the protein-bound label. If the value is higher than this, perform another 1:200 buffer exchange into PBSZ with a new centrifugal concentrator unit.

9. Determine the conjugation efficiency from the molar ratio of the conjugated label to protein, calculated from the absorbance of the sample at the fluorophore λ_{max} and at 280 nm. These measurements are made with a spectrophotometer.

$$\text{Efficiency} = \frac{\text{Abs}_{\lambda_{max}}}{\text{Ext}_{label}} \times \frac{\text{Ext}_{A1AT}}{\text{Abs}_{280\ nm} - \text{Abs}_{\lambda_{max}} * p}$$

Ext_{label} is the extinction coefficient of the label at λ_{max}, p is the ratio of the absorbance of the label at 280 nm of its absorbance at λ_{max} (both parameters should be provided by the manufacturer), Ext_{A1AT} is 2.7×10^4 cm^{-1} M^{-1} for plasma-derived A1AT, and 2.6×10^4 cm^{-1} M^{-1} for recombinant A1AT.

10. A poor labeling efficiency will disproportionately impact the intensity of the FRET signal. Therefore, for the experiments described here, the labeling process should be repeated with freshly reduced material if the calculated value is below 0.7.

11. Concentrate the labeled protein to around 1–2 mg ml^{-1}, divide it into aliquots, and freeze it at −80 °C until it needs to be used.

3.2 Sample Preparation Using Anion Exchange Chromatography

This affinity chromatography step is useful for separating the A1AT preparation from contaminants such as reducing agent or unconjugated fluorophore. Size exclusion chromatography can also be used, but has not been found to be as effective in the latter instance.

1. Connect a 1 ml or 5 ml Q sepharose column to a chromatography system, "purge" the tubing and pump with ultrapure water, and pass three column volumes of water through the column at the manufacturer's recommended flow rate.

2. Place the inlets of the chromatography system into anion exchange buffers A and B, and "purge" the system to ensure that all of the tubing contains the appropriate buffer. Wash the column with three column volumes of buffer B (or until any residual material from previous experiments is eluted, as monitored by the absorbance at 280 nm), and then equilibrate it into buffer A.

3. Dilute the A1AT preparation into anion exchange buffer A in order to ensure an appropriate pH and low salt concentration. For samples in PBSZ, a 1:3 dilution is sufficient.

4. Apply the diluted sample to the column.

5. Remove the unbound sample and buffer components (such as the reducing agent) by using ten column volumes of anion exchange buffer A.

6. Apply a salt gradient to the column, from 0 to 50% buffer B across 16 column volumes. Use the fraction collector from the chromatography system to collect half-column volume fraction sizes of the eluted material. The profile of the absorbance at 280 nm can be used to identify the fractions of interest (*see* **Note 10**).

7. Wash the chromatography system and Q sepharose column with ultrapure water, followed by 20% ethanol in preparation for storage.

3.3 Heat-Induced Polymerization Monitored by FRET

Typically, depending on the presence of other destabilizing factors such as mutations or denaturant, heating A1AT to a temperature in the range of 50 °C–65 °C results in the formation of polymers in a matter of minutes to hours. When the polymers contain subunits with fluorophores whose fluorescence profiles partially overlap, FRET can be used to monitor this process. Real-time thermal cyclers are ideally suited for this application, due to accurate temperature control, small sample requirements, and the ability of some thermal cyclers to generate temperature gradients across a plate.

1. Select a real-time PCR plate that is suitable for use with the chosen thermal cycler. Design a plate layout that is appropriate to the experimental aims. For example, the experiment may explore the effect of different temperatures, A1AT concentrations, or additives on the rate of polymerization. If a thermal cycler is used that is capable of producing a temperature gradient across the plate, the sample composition might be kept constant across columns and vary down rows.

2. Based on the conjugation efficiencies of the two fluorescent A1AT preparations, mix the samples such that the labels are in an equal ratio.

3. Dilute the combined sample to the working concentration in the chosen buffer, and dispense 20 µl aliquots into wells. Based on previous experience, a concentration between 50 µg ml^{-1} and 200 µg ml^{-1} is a good starting place.

4. Seal the plate with transparent film that is suitable for real-time thermal cyclers, and centrifuge the plate at $500 \times g$ for 1 min.

5. If necessary, calibrate the thermal cycler for the individual fluorophores according to the instrument protocol (*see* **Note 11**). Create a constant-temperature (non-temperature cycling) program of a suitable length for the thermal cycler, ensuring that the fluorescence is read at appropriate intervals (ideally a minimum of every 30 s). A heated lid should be used, as sample evaporation will affect the results. Heating and cooling rates should be set at their maximum. The fluorescence both of the donor and acceptor should be recorded.

6. Perform the polymerization experiment.

7. Calculate the relative FRET efficiency by dividing the fluorescence of the acceptor by that of the donor for each time point, using an appropriate numerical package. This ratiometric calculation can reduce the impact of anomalies such as small bubbles. The resulting progress curves typically show a sigmoidal increase, followed by a gradual decline in fluorescence (Fig. 1). The half-time of polymerization, the time at which the increasing signal reaches the midpoint between the maximum and minimum of the curve, is a convenient way to represent the rate of the reaction. This can be readily determined by scaling the FRET data to fall between 0 and 1, and reading the time at which the value reaches 0.5.

3.4 End-Point Experiments of Polymerization Induced at Multiple Temperatures

The susceptibility of an A1AT variant to polymerization can be readily and rapidly assessed by incubating the material over a temperature range for a fixed period of time and visualizing the result by native PAGE (Fig. 2). This can be useful to obtain a qualitative comparison between variants, overcome sample

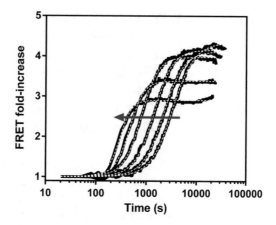

Fig. 1 *The polymerization of A1AT as reflected by an increase in FRET.* The *graph* shows the progress curves obtained during the polymerization of fluorescently labeled A1AT in a real-time thermal cycler. The curves were normalized by reporting fluorescence as a fold-increase over the starting FRET value. A temperature gradient was generated across the plate. The rate of polymerization increases with increasing temperature, as indicated by the *arrow*

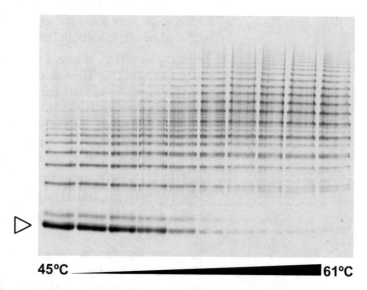

Fig. 2 *Polymerization end-point experiment.* Plasma-purified M A1AT was incubated for 18 h over a 45.0–60.7 °C temperature range in PBS containing 5% glycerol. The resulting polymers were resolved with a native PAGE and the protein was visualized with Coomassie *blue* stain. The position of monomer is indicated by a *triangle*

heterogeneity (when coupled with western blot analysis), or to determine appropriate conditions to undertake preparative polymerization.

1. Dilute A1AT, either plasma-derived or recombinant, to 200 μg ml^{-1} in a suitable buffer, for example PBSZ.

2. Dispense 20 μl sample aliquots into wells of a PCR plate (or PCR tubes) spanning the width of a thermal cycler heating block.

3. Cap the tubes or seal the plate. If necessary, centrifuge briefly at $500 \times g$ to ensure that the samples are at the bottom of the wells.

4. Program the thermal cycler to incubate the samples with a single fixed-time step, consisting of a temperature gradient across the plate, for example between 45 °C and 60°C. Durations might be 2, 4, or 18 h. Conclude with an indefinite 12 °C step.

5. Insert the tubes or plate into the thermal cycler and perform the incubation using a heated lid to prevent evaporation.

6. The thermal cycler will have a display that reports the actual temperatures across the sample block. Take note of these temperatures, as the distribution of values will not be linear.

7. Add 5 μl of native loading buffer to each well or tube. In addition, prepare a monomer control: 20 μl of 200 μg ml^{-1} with 5 μl of native loading buffer.

8. The samples will be resolved by clear native PAGE. Commercially available Tris/Glycine or bis-Tris-based polyacrylamide gel systems work well, but Tris-Borate-EDTA (TBE) polyacrylamide gels do not. If using a system such as NativePAGE™ (Life Technologies), *do not* use the tank dye additive that is used for blue native PAGE applications. Set up the gel running apparatus as described in the associated protocol.

9. Load the gel with the monomer control, and heated samples. If using a 15-well gel, add 15 μl of each rather than the full 25 μl.

10. Electrophoresis is performed at 20–25 mA of constant current for a single gel.

11. The protein can be visualized by using a Coomassie blue-based stain, SYPRO Orange, or using a UV transilluminator if the material has been prelabeled with a fluorophore. As an alternative, a western blot can be performed using an anti-A1AT antibody.

3.5 Polyacrylamide-Agarose Hybrid "Slab" Gels

Polyacrylamide gels can be cast in a horizontal slab format to resolve proteins with an apparatus typically used for DNA electrophoresis. A larger number of samples can be processed than typically possible with vertical mini-gel systems, with around 60 wells for a 15 × 20 cm gel format (Fig. 3). The separation of the A1AT polymers depends on the molecular weight and charge, and the use of an acrylamide-agarose mix ensures a strong gel matrix. Amounts

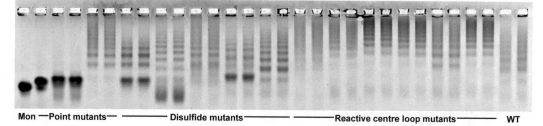

Mon ⟞Point mutants⟞ ⟞————— Disulfide mutants—————⟞ ⟞————Reactive centre loop mutants————⟞ WT

Fig. 3 *Slab gel electrophoresis of polymers.* The top half of an acrylamide-agarose slab gel, illustrating the ability to process a large number of samples at once. Different recombinant A1AT mutants were subjected to heat-induced polymerization at 50 °C for 18 h, and resolved in duplicate according to the method described here. The outermost left lane shows monomeric recombinant A1AT, and the two outermost right lanes show the wild-type polymer (labeled "Mon")

reported are for a 100 ml gel volume. Work in a fume hood, and ensure that all of the casting components are readily available as the gel mixture sets rapidly.

1. Place the gel mold in the casting case.

2. In a 250 ml conical flask, add 16.7 ml of a 30% acrylamide/bis-acrylamide (37.5:1) solution, 10 ml of 10× cathode buffer, 22 ml ultrapure water, and 80 μl of TEMED. Swirl gently.

3. In a separate conical flask prepare 50 ml of a 2% agarose solution in ultrapure water, and boil in a microwave until it is dissolved. The solution should be monitored while heating to prevent it from boiling over.

4. Allow the agarose to cool slightly before pouring the acrylamide solution into the agarose in a smooth motion, gently swirling until it is mixed.

5. Add 1 ml of 10% ammonium persulfate, swirling gently, and quickly transfer the solution to the gel mold. Insert the combs to create the desired number of wells.

6. Allow the gel to set and cool, before placing in the electrophoresis tank. Fill the tank with 1× cathode running buffer so that the gel is submerged. For best results, run the gel at 4 °C.

7. Add loading buffer to the protein samples, and carefully pipette them into the wells.

8. Run the gel at a constant 100–150 V for 2–2.5 h.

9. Gels can be stained with Coomassie blue for 2–4 h followed by destaining. Alternatively, SYPRO Orange™ is a more effective stain for this gel format when used as described in **steps 10–14**.

10. Incubate the gel in 7.5% acetic acid and 0.1% SDS for 2 h at room temperature.

11. Rinse twice with 7.5% acetic acid to remove the SDS.

12. Place the gel in a solution of 7.5% acetic acid containing 1:5000 SYPRO Orange™ overnight at room temperature, in the dark.

13. Rinse the gel with 7.5% acetic acid for 30 min.

14. Record using a fluorescence scanner with an excitation of around 490 nm and an emission reading in the 590–610 nm range.

4 Notes

1. This step should be done quickly as conjugates are inactivated by oxidation. In our experience, reactivity is maintained for at least 6 months when prepared in this way. Real-time thermal cycler instruments usually excite around 470 nm, so a fluorescein-like "donor" is used, such as Alexa Fluor™-488 (Life Technologies) or Atto™-488 (AttoTec). The "acceptor" should have an overlap in excitation spectrum with the emission spectrum appropriate for the periodicity of the polymer around 70–80 Å. Alexa Fluor™-594 (Life Technologies) and Atto™-594 (AttoTec) work well in this regard.

2. Plasma-derived A1AT can be obtained as a pharmaceutical preparation, as a research reagent, or can be conveniently purified using commercially available affinity resin (Antitrypsin Select, GE Biosciences).

3. During purification of recombinant or wild-type A1AT, it can be convenient to reduce the endogenous cysteine at position 232 in preparation for later conjugation. Typically, the final step makes use of a Q sepharose resin or gel filtration column. If 10 mM dithiothreitol is added to the sample immediately prior to the application on the columns, the purification will effectively remove the excess reducing agent. The sample should be frozen soon after the elution.

4. If a chromatography system is not available, a "gravity feed" method can be used in which loose resin at a final volume of 1–5 ml is deposited into a column with a sintered glass frit, whose volume is several times that of the resin.

5. Buffers can be easily prepared from stock solutions of 1 M Tris pH 8.0, 5 M NaCl, and 20% sodium azide which have been filtered. If the ultrapure water is delivered through an inline filter (typical in modern systems), buffers can be prepared by mixing these components without filtering.

6. Some real-time thermal cyclers use detectors/filters that individually span very wide wavelength ranges. It is important that

the fluorescence emission of the donor and that of the acceptor do not fall into the same detection "bin."

7. Fluorescein derivatives generally have an excitation spectrum that extends into the near UV region, and thus can be readily visualized with a UV transilluminator. The low background and linear response makes this a useful approach when densitometry is used.

8. Acrylamide is a neurotoxin with a cumulative effect. Therefore, care must be exercised when handling and disposing of this chemical.

9. Fluorophores can be susceptible to photobleaching, although modern variants tend to be relatively photostable, provided they are not left unnecessarily exposed to light. For previously conducted experiments, it was not necessary to aliquot or mix the fluorophores in a darkened room. However, reactions and storage should be undertaken in the dark, for example by covering the tubes with foil. Similarly, chromatography resins and eluted fractions should be shaded from the light.

10. As fluorophores can often absorb light at 280 nm, resolve the fractions by SDS-PAGE and visualize the gel with a UV transilluminator or fluorescence gel scanner (without staining). Unconjugated fluorophores will tend to run near the dye front.

11. If it is possible to export raw (non-deconvoluted) fluorescence data, this is generally preferred.

Acknowledgments

A.E.B. is supported by a Rosetrees Trust PhD studentship grant. I.H. was an eALTA Fellow. This work was funded in part by a grant from the Alpha-1 Foundation to J.A.I.

References

1. Gooptu B, Lomas DA (2009) Conformational pathology of the serpins—themes, variations and therapeutic strategies. Annu Rev Biochem 78:147–176

2. Lomas DA, Evans DL, Finch JT, Carrell RW (1992) The mechanism of Z alpha 1-antitrypsin accumulation in the liver. Nature 357:605–607

3. Lomas DA, Finch JT, Seyama K, Nukiwa T, Carrell RW (1993) Alpha 1-antitrypsin Siiyama (Ser53->Phe). Further evidence for intracellular loop-sheet polymerization. j Biol Chem 268:15333–15335

4. Berg NO, Eriksson S (1972) Liver disease in adults with alpha-1 -antitrypsin deficiency. N Engl J Med 287:1264–1267

5. Ordóñez A, Snapp EL, Tan L, Miranda E, Marciniak SJ, Lomas DA (2013) Endoplasmic reticulum polymers impair luminal protein mobility and sensitize to cellular stress in alpha1-antitrypsin deficiency. Hepatology 57:2049–2060

6. Hidvegi T, Schmidt BZ, Hale P, Perlmutter DH (2005) Accumulation of mutant alpha1-antitrypsin Z in the endoplasmic reticulum activates caspases-4 and -12, NFkappaB, and BAP31 but not the unfolded protein response. J Biol Chem 280:39002–39015

7. Kröger H, Miranda E, MacLeod I, Pérez J, Crowther DC, Marciniak SJ, Lomas DA (2009) Endoplasmic reticulum-associated degradation (ERAD) and autophagy cooperate to

degrade polymerogenic mutant serpins. J Biol Chem 284:22793–22802

8. Bruch M, Weiss V, Engel J (1988) Plasma serine proteinase inhibitors (serpins) exhibit major conformational changes and a large increase in conformational stability upon cleavage of their reactive sites. J Biol Chem 263:16626–16630

9. Carrell RW, Owen MC (1985) Plakalbumin, a1-antitrypsin, antithrombin and the mechanism of inflammatory thrombosis. Nature 317:730–732

10. Yamasaki M, Li W, Johnson DJ, Huntington JA (2008) Crystal structure of a stable dimer reveals the molecular basis of serpin polymerization. Nature 455:1255–1258

11. Yamasaki M, Sendall TJ, Pearce MC, Whisstock JC, Huntington JA (2011) Molecular basis of alpha1-antitrypsin deficiency revealed by the structure of a domain-swapped trimer. EMBO Rep 12:1011–1017

12. Mast AE, Enghild JJ, Salvesen G (1992) Conformation of the reactive site loop of alpha 1-proteinase inhibitor probed by limited proteolysis. Biochemistry 31:2720–2728

13. Devlin GL, Chow MK, Howlett GJ, Bottomley SP (2002) Acid denaturation of alpha1-antitrypsin: characterization of a novel mechanism of serpin polymerization. J Mol Biol 324:859–870

14. Ekeowa UI, Freeke J, Miranda E, Gooptu B, Bush MF, Perez J, Teckman J, Robinson CV, Lomas DA (2010) Defining the mechanism of polymerization in the serpinopathies. Proc Natl Acad Sci U S A 107:17146–17151

15. Miranda E, Perez J, Ekeowa UI, Hadzic N, Kalsheker N, Gooptu B, Portmann B, Belorgey D, Hill M, Chambers S, Teckman J, Alexander GJ, Marciniak SJ, Lomas DA (2010) A novel monoclonal antibody to characterize pathogenic polymers in liver disease associated with alpha1-antitrypsin deficiency. Hepatology 52:1078–1088

16. Powell LM, Pain RH (1992) Effects of glycosylation on the folding and stability of human, recombinant and cleaved a1-antitrypsin. J Mol Biol 224:241–252

17. Dafforn TR, Mahadeva R, Elliott PR, Sivasothy P, Lomas DA (1999) A kinetic mechanism for the polymerization of alpha1-antitrypsin. J Biol Chem 274:9548–9555

18. Irving JA, Miranda E, Haq I, Perez J, Kotov VR, Faull SV, Motamedi-Shad N, Lomas DA (2015) An antibody raised against a pathogenic serpin variant induces mutant-like behaviour in the wild-type protein. Biochem J 468:99–108

19. Haq I, Irving JA, Faull SV, Dickens JA, Ordonez A, Belorgey D, Gooptu B, Lomas DA (2013) Reactive centre loop mutants of alpha-1-antitrypsin reveal position-specific effects on intermediate formation along the polymerization pathway. Biosci Rep 33:e00046

20. Tsutsui Y, Dela Cruz R, Wintrode PL (2012) Folding mechanism of the metastable serpin alpha1-antitrypsin. Proc Natl Acad Sci U S A 109:4467–4472

21. Tsutsui Y, Kuri B, Sengupta T, Wintrode PL (2008) The structural basis of serpin polymerization studied by hydrogen/deuterium exchange and mass spectrometry. J Biol Chem 283:30804–30811

22. Mallya M, Phillips RL, Saldanha SA, Gooptu B, Brown SC, Termine DJ, Shirvani AM, Wu Y, Sifers RN, Abagyan R, Lomas DA (2007) Small molecules block the polymerization of Z alpha1-antitrypsin and increase the clearance of intracellular aggregates. J Med Chem 50:5357–5363

23. Ordóñez A, Pérez J, Tan L, Dickens JA, Motamedi-Shad N, Irving JA, Haq I, Ekeowa U, Marciniak SJ, Miranda E, Lomas DA (2015) A single-chain variable fragment intrabody prevents intracellular polymerization of Z α1-antitrypsin while allowing its antiproteinase activity. FASEB J 29:2667–2678

Chapter 25

Therapeutics: Alpha-1 Antitrypsin Augmentation Therapy

Michael Campos and Jorge Lascano

Abstract

Subjects with alpha-1 antitrypsin deficiency who develop pulmonary disease are managed following general treatment guidelines, including disease management interventions. In addition, administration of intravenous infusions of alpha-1 proteinase inhibitor (augmentation therapy) at regular schedules is a specific therapy for individuals with AATD with pulmonary involvement.

This chapter summarizes the manufacturing differences of commercially available formulations and the available evidence of the effects of augmentation therapy. Biologically, there is clear evidence of in vivo local antiprotease effects in the lung and systemic immunomodulatory effects. Clinically, there is cumulative evidence of slowing lung function decline and emphysema progression. The optimal dose of augmentation therapy is being revised as well as more individualized assessment of who needs this therapy.

Key words Alpha-1 antitrypsin deficiency, Alpha-1 proteinase inhibitor, COPD, Emphysema, Clinical trials, Augmentation therapy

1 Introduction

Subjects with alpha-1 antitrypsin deficiency (AATD) who develop pulmonary disease (chronic obstructive pulmonary disease, COPD) are managed following general COPD treatment guidelines. These nonspecific therapies include administration of inhaled bronchodilators, inhaled steroids, pulmonary rehabilitation, smoking cessation vaccination against respiratory pathogens, aggressive treatment of acute exacerbations, and oxygen therapy and lung transplantation for selected individuals. These interventions decrease the morbidity and mortality of COPD and can be integrated through educative efforts as part of a disease management program to significantly improve the health-related quality of life in AATD [1]. In addition, administration of intravenous infusions of alpha-1 proteinase inhibitor (a1-PI) at regular schedules is a specific therapy for individuals with AATD with pulmonary involvement. This "augmentation therapy" originated more than 30 years ago from our understanding of the pathophysiology of lung disease in AATD with the rationale to correct the deficient

Florie Borel and Christian Mueller (eds.), *Alpha-1 Antitrypsin Deficiency: Methods and Protocols*, Methods in Molecular Biology, vol. 1639, DOI 10.1007/978-1-4939-7163-3_25, © Springer Science+Business Media LLC 2017

state of alpha-1 antitrypsin (AAT) in serum and lung tissues [2]. Since then, a growing amount of evidence has accumulated to justify its use in this condition.

1.1 Manufacturing Process and Available Products

Because of its therapeutic utility, commercial production of a1-PI has been the subject of considerable research. Although progress has been made in the production of recombinant a1-PI in *E. coli* [3], yeast [4], plants [5] and by secretion in the milk of transgenic mammals [6, 7], isolation of alpha-1 antitrypsin (AAT) from human plasma is presently the most efficient practical method of obtaining the protein in sufficient quantities. Therefore, human plasma is the only FDA-approved source and all commercially available a1-PI preparations derive from pooled human plasma. The six commercial a1-PI preparations that have received FDA approval in the United States so far contain slight variations in purification methods (Table 1). Most published processes for a1-PI isolation begin with one or more fractions of human plasma known as the Cohn fraction IV precipitates, which are obtained from plasma as a paste after a series of ethanol precipitations and pH adjustments [8]. The following steps for isolating and purifying the protein involve combinations of precipitation, adsorption, extraction, and chromatographic steps. The resulting product should follow certain standards that include: less than 6% (ideally less than 2% and most preferably less than 1%) of contaminating serum proteins; an apparent ratio of active to antigenic a1-PI greater than 1.08 (preferably greater than 1.16 and most preferably greater than 1.23) as measured by nephelometry; and a reduction of enveloped viruses by at least 11 $\log_{10}$ units, and non-enveloped viruses by at least 6 $\log_{10}$ units when measured in in-vitro spiking studies using human viruses. In addition, the product should be stable for at least 2 years when stored lyophilized at up to 25 ° C.

The initial clinical augmentation therapy trials in subjects with AATD were conducted in the early 1980s. Without data regarding which was the lowest protective serum AAT level, it was estimated that a target of 11 μM was a reasonable goal, based on the observation that individuals with genotype SZ that never smoked and had AAT levels above this threshold rarely developed lung disease [9]. Wewers et al. conducted the pivotal trial targeted to achieve serum AAT levels above 11 μM with dosing and frequency tested in three phases [10]. It concluded that weekly administration of 60 mg/kg was safe and effective in maintaining serum levels and bronchoalveolar lavage levels above the predicted thresholds.

Based on these results, the FDA approved augmentation therapy (Prolastin®) (Grifols, Research Triangle Park, NC) in 1987 supported by the Orphan Drug Act with the requirement to establish a patient registry program to follow and monitor 1000 subjects for long-term safety with comparison with nontreated individuals. This was followed by the approval for Aralast® (Baxter Healthcare,

Table 1

Comparison of a1-PI commercial preparations available in the United States

| | FDA approval year | Purification method | | | | | | | | Needs reconstitution | Concentration after reconstitution | Infusion time for 60 mg/kg dose (min) |
		Cold ethanol fractionation	PEG precipitation	Zn Cl precipitation	Chromatography	Depth filtration	Pasteurization	Solvent detergent purification	Nano filtration			
Prolastin®	1987	√	√			√	√			Yes	25 mg/ml	30'
Aralast®	2002	√	√	√	√			√	√	Yes	20 mg/ml	37.5'
Zemaira®	2003	√					√		√	Yes	50 mg/ml	15'
Aralast NP®a	2007	√	√	√	√			√	√	Yes	20 mg/ml	37.5'
Prolastin-C®b	2009	√	√		√	√		√	√	Yes	50 mg/ml	15'
Glassia®	2010	√			√			√	√	No	20 mg/ml	60–80'

[a]Compared to Aralast: C-terminal lysine (lys 394) was removed

[b]Compared to Prolastin: More purified, more concentrated and faster infusion time

Deerfield IL) in 2002, Zemaira® (CSL Behring, Kankakee, IL) in 2003, Aralast NP® (Baxter Healthcare, Deerfield IL) in 2007, Prolastin-C® (Grifols, Research Triangle Park, NC) in 2009, and Glassia® (Kamada, Beit Kama, Israel) in 2010.

1.2 Biological Evidence: Neutrophil Elastase Inhibition

Augmentation therapy augments AAT levels in bronchoalveolar lavage and increases the anti-elastase capacity of serum and epithelial lining fluid [10, 11]. The recommended dosing regimen of 60 mg/kg/week leads to a 60–70% increase in the epithelial lining fluid's capacity to inhibit neutrophil elastase (NE) activity [10].

To test its efficacy in slowing emphysema progression, several markers of elastin degradation have been evaluated. Two of these markers, desmosine and isodesmosine, can be measured in urine, serum, or bronchoalveloar lavage and have been found to be more elevated in subjects with AATD [12, 13]. Comparisons of desmosine and isodesmosine levels between patients receiving augmentation or not and comparisons of these biomarkers before and after receiving augmentation therapy, conclude that AAT replacement confers protection against elastin degradation [14]. Another recently studied plasma biomarker is Aα-Val^{360}, a specific cleavage product generated by the action of NE on fibrinogen and used as a specific surrogate marker of pre-inhibition NE activity. Aα-Val^{360} correlates cross-sectionally to physiologic, radiologic, and symptomatic markers of disease severity in AATD [15] and demonstrates a treatment response in individuals who receive augmentation therapy [16].

1.3 Biological Evidence: Anti-Inflammatory Effects

In recent years, it has been observed that AAT has physiological effects other than protease inhibition, including cellular protection against apoptosis and modulation of inflammation [17, 18]. AAT has an inhibitory effect over neutrophil chemotaxis and modulates TNF-alpha production and synthesis from neutrophils, effects that are lost in subjects with AATD with a subsequent increase in the production of inflammatory markers. It has been shown that subjects with AATD show significant increases in neutrophil counts and inflammatory markers (IL-8, IL-6, and IL-1β) in bronchoalveolar lavage fluid when compared to subjects without AATD [19], even with only mild reductions in lung function. This neutrophil "hyperactivity" can be mitigated in vivo with the administration of augmentation therapy [20, 21]. Within a week, augmentation therapy can lower serum IL-8 levels [22].

The effect of augmentation therapy on airway inflammatory markers was evaluated by Stockley et al. in 12 subjects with AATD (60 mg/kg weekly for 4 weeks) [23]. Compared to pretreatment values, augmentation therapy led to a rise of AAT concentrations in sputum similar to those of nondeficient subjects. Furthermore, the treatment was associated with a favorable change in the sputum inflammatory milieu, with statistical significant reductions in sputum elastase activity and levels of the chemoattractant leukotriene B4, and a trend toward decreased sputum myeloperoxidase and IL-8.

1.4 Clinical Evidence: Slowing Lung Function Decline

Over the last 30 years, several nonrandomized observational studies focused on the effect of augmentation therapy on lung disease progression. In 1997, a study showed that 97 Dutch subjects not receiving augmentation therapy had a significantly greater FEV_1 decline compared to 198 German subjects on therapy after 4.5 years of follow-up (FEV_1 = 53 ml/year versus 75 ml/year, respectively, $p = 0.02$) [24]. A year later the results from the NHLBI Alpha-1 Registry ($N = 1129$) failed to show an overall statistically significant difference in FEV_1 decline. However in the group of subjects with a moderate reduction in lung function (mean FEV_1 35 to 49% predicted), subjects on augmentation therapy exhibited slower lung function decline (mean difference = 27 ml/year, 95% CI: 3–51 ml/year; $p = 0.03$) [25]. A third study investigated the rate of lung function decline before and after the initiation of augmentation therapy in a cohort of 96 German patients with AATD [26]. The study showed that the decline in FEV_1 was significantly slower during the treatment period versus the pretreatment period (34.3 ml/year versus 49.2 ml/year, $p = 0.019$). Finally, a retrospective observational study using data collected via the Canadian AIR Registry observed a slower decline in 21 subjects receiving augmentation therapy compared to 42 untreated controls followed for 5.6 years (29.9 ml/year versus 63.6 ml/year, $p = 0.019$) [27]. A subsequent meta-analysis of these and other studies reported an overall significantly slower decline in FEV_1 among subjects that received augmentation therapy when compared to nontreated controls, with an effect particularly significant among subjects with moderate to severe reductions in lung function (30–65% predicted) [28].

The observational studies provided useful information for the design of subsequent prospective placebo-controlled studies. For example, based on the NHLBI Registry study data, to detect a difference in FEV_1 decline of 23 ml/yr. (i.e., a 28% reduction) a study would require enrollment of 147 subjects with moderate to severe lung function impairment per treatment arm over a 4-year period, while a study to detect a 40% reduction in mortality would require 342 subjects per treatment arm over a 5-year period [29]. These calculations highlight the need to use more efficient clinical markers of disease progression to assess the clinical effects of augmentation therapy.

1.5 Clinical Evidence: Slowing Radiologic Progression of Emphysema

Changes lung density measurements by computed tomography (CT) densitometry have been shown to be a sensitive measure for assessing emphysema progression and as an outcome measure of emphysema-modifying therapy in patients with AATD [30]. In addition, CT emphysema measures correlate with FEV_1 decline, health status, exercise capacity, and even mortality in this population [31, 32].

To date, all trials using CT densitometry as an outcome have been randomized trials using albumin infusions as a comparator. In the first of these trials, Dirksen et al. evaluated the effect of 250 mg/kg of a1-PI administered at 4 week intervals for at least 3 years in 56 patients with moderate lung impairment (FEV_1 30–80% predicted). Admittedly an underpowered study, it showed a trend for a reduction in the loss of lung tissue in the group who received augmentation therapy compared to the placebo group [33]. The EXAcerbations and CT scan as Lung Endpoints (EXAC-TLE) trial studied the effect of 60 mg/kg a1-PI administered weekly in 77 subjects for up to 30 months. Also an underpowered study, the trial included several statistical analyses, all which suggested at least a trend toward a positive effect of augmentation therapy in reducing loss of lung density (with p-values ranging from 0.049 to 0.084) [34]. Two published meta-analyses of these two trials have favored augmentation therapy as effective in slowing emphysema progression assessed by CT densitometry [35, 36].

The more recent multicenter Randomized, placebo-controlled trial in Alpha-1 Proteinase Inhibitor Deficiency (RAPID) trial studied the effect of 60 mg/kg/week of a1-PI in 92 subjects compared to 85 receiving placebo over 2 years. The study showed that the annual rate of lung density loss by CT scan when measured at total lung capacity was slower in the treatment group ($p = 0.03$) [37]. Interestingly, in an open label 2-year study extension, the rate of lung density loss that was greater in patients who were taking placebo during the double-blind portion of the trial slowed to parallel that of patients who had received active treatment throughout in the extension study [38]. Altogether, the results of these randomized studies show that intravenous augmentation slows emphysema progression by decreasing loss of lung density as measured by CT imaging.

A summary of published clinical trials has been summarized in Table 2.

1.6 Clinical Evidence: Other Outcomes

Augmentation therapy has not been shown to modify other clinical outcomes commonly measured in COPD trials such as reducing exacerbation frequency or mortality. An initial web-based patient survey conducted by Lieberman et al. suggested a possible beneficial effect of augmentation therapy on exacerbations [39]. However, other prospective studies have shown that acute exacerbations

Table 2
Summary of studies addressing the clinical effects of augmentation therapy

	Group	Augmentation therapy group	Nontreated group	P
Observational studies: FEV$_1$ decline as outcome				
Seersholm et al. [24] (N = 97)	ALL	53 ml/year	75 ml/year	0.02
	Subjects with FEV$_1$ 31%–65%	62 ml/year	83 ml/year	0.04
NHLBI registry [25] (N = 1129)	ALL	51 ml/year	56 ml/year	NS
	Subjects with FEV$_1$ 35%–49%	66.4 ml/year	93.2 ml/year	0.03
Wencker et al. [26] (pre-post study, N = 96)	ALL	34.2 ml/year	49.2 ml/year	0.019
	Subjects with FEV$_1$ 30%–65%	37.8 ml/year	49.3 ml/year	NS
	Subjects with FEV$_1$ > 65%	48.9 ml/year	122.5 ml/year	0.001
Chapman et al. [27] (N = 42)	ALL	30 ml/year	63 ml/year	0.019

Randomized controlled studies: change in CT lung density as outcome

	Analysis	Augmentation—Placebo treatment difference (g/L/year)	P
Dirksen et al. [33] (monthly treatment for 4 years, N = 56)	15th percentile lung density (PD15) adjusted for lung volume (g/L)	8.9 (2.6–11.5)	0.07
EXACTLE trial [34] (weekly treatment for 2–2.5 years)	Method 1 (physiological, slope analysis)	0.8 (−0.065–1.778)	0.068
	Method 2 (statistical, slope analysis)	0.7 (−0.028–1.427)	0.059
	Method 3 (physiological, end-point analysis)	1.6 (−0.220–3.412)	0.084
	Method 4 (statistical, end-point analysis)	1.5 (0.009–2.935)	0.049
RAPID trial [37] (weekly treatment for 2 years, N = 180 and a 2-year extension trial, N = 97)	Annual rate of lung density loss at TLC and FRC combined	0.62 (0.02–1.26)	0.06
	annual rate of lung density loss at TLC	0.74 (0.06–1.42)	0.03
	annual rate of lung density loss at FRC	0.048 (0.22–1.18)	0.18

are common in subjects with AATD [40, 41] and randomized trials have not shown a trend toward exacerbation number reduction [34, 37].

The NHLBI Registry Group study reported a decrease in mortality among subjects receiving augmentation therapy, an effect particularly noted for subjects with FEV_1 below 50% (risk ratio [RR] = 0.64, 95% CI: 0.43–0.94, p = 0.02); however, its non-randomized nature and regional differences in patient management prompt caution when interpreting these results [25].

1.7 Augmentation Therapy: Safety and Tolerability

The experience of thousands of treated individuals suggests that augmentation therapy is generally safe and well tolerated. The few side effects reported were generally mild and rarely required major interventions or interruption of therapy.

Wencker et al. reported the side effects associated a total of 58,000 a1-PI infusions administered to 443 subjects over 6 years, with only 124 reported side effects most commonly fever/chills, urticarial, nausea and vomiting, and fatigue [42]. There were only five severe side effects reported during the study that required medical intervention or hospitalizations, including four patients who suffered an anaphylactic reaction. No deaths or viral transmission related to augmentation therapy was observed during the study period. Safety data from the NHLBI Registry (n = 747) showed a low overall rate of adverse events at 0.02 per patient-month, with 83% of patients reporting no events [43]. Again, the most common complaints were headache (47%), dizziness (17%), nausea (9%), and dyspnea (9%), the latter being classified as severe. No transmission of viral hepatitis, HIV infection, or prion disease was reported. A rare case of an IgE-mediated anaphylactic reaction following the third intravenous infusion of the original formulation of Prolastin® has been reported [44].

1.8 Augmentation Therapy: Cost-Effectiveness

Augmentation therapy is expensive, with an average annual total health-care cost ranging from $36,471 to $46,114 [45], with the use of augmentation accounting for more than half of all healthcare costs for an AATD individual [46]. However, some estimates increase the cost to be as high as US$120,000 annually, thrice the yearly cost for patients with COPD alone [47]. Nevertheless, it is estimated that augmentation therapy increases years of life gained at a cost comparable to that of other evidenced-based interventions [48]. Specifically, female smokers were estimated to gain a mean of 7.14 years, female nonsmokers 9.19 years, male smokers 5.93 years, and male non-smokers 10.60 years. Alkins and O'Malley calculated the incremental cost (the value added to usual care) per year of life saved by augmentation therapy to be $13,971 and concluded it to be a cost-effective intervention [49]. But using Monte Carlo simulation, Gildea et al. suggested that the cost of augmentation therapy would need to be reduced from US$54,765 to US$4900 to be deemed a lifelong cost-effective therapy [50].

Data from registries have shown a great variability in the evolution of lung disease in AATD. Since avoidance of risk factors may stabilize disease progression in many cases, some have advocated a personalized approach to treatment in order to increase the cost-effectiveness of augmentation therapy, considering a combination of age, physiological impairment, exacerbation history, and rate of lung function decline by spirometry and other measures of emphysema [51].

1.9 Augmentation Therapy: Treatment Recommendations and Considerations

Current standard of care guidelines recommend augmentation therapy for individuals with abnormal AAT genotypes who have serum AAT levels below 11 μM and have evidence of airflow obstruction [2]. The FDA-approved dosing is of 60 mg/kg administered intravenously once a week. Heterozygous individuals (i.e., PI MZ or MS) usually have serum AAT levels above 11 μM, and although with levels below the normal range, current recommendations do not endorse use of augmentation therapy for such individuals due to lack of clinical studies [52].

However, in clinical practice variability in prescription patterns are frequently observed. In the NHLBI study, 25.3% of participants received biweekly and 21.8% received monthly infusions at the beginning of the study with additional patients switching to non-weekly regimens during the study period [25]. In more recent years, 32.4% of 922 US patients received infusions every 2 weeks and 7.6% received monthly infusions [53]. It is clear that these non-weekly regimens will not complete confer adequate AAT serum levels throughout the dosing intervals. Dosing a1-PI at 250 mg/kg every 28 days confers protective serum AAT levels and anti-elastase activity in epithelial lining fluid for "at least" 25 days after the infusion [54], while a regimen of 120 mg/kg every 2 weeks could not maintain nadir serum levels above the threshold for the entire 14-day dosing interval [55]. Using a tri-compartment pharmacokinetic model it has been suggested that weekly doses of 50 mg/kg or bi-weekly doses of 120 mg/kg may confer protective trough levels at least 85% of the time, while a monthly regimen is not suitable as it protected for only 22 of the 28 days [56]. The factors that may be associated with increased clearance of a1-PI have not been defined; however, it is known that independent variables such as sex, age, and body weight have no influence on pharmacokinetic parameters [57].

1.10 Augmentation Therapy: Future Considerations

Several new studies and observations are currently challenging the future landscape of augmentation therapy as is currently recommended. These include newer dosing recommendations, alternate delivery routes, and genetic therapy.

The therapeutic target to augment trough serum AAT levels over 11 μM has been challenged given the difficulty to prove the clinical benefits of the current recommended dosing of 60 mg/kg/

week and the fact that AAT levels in nondeficient individuals are twice as high (22–50 µM). Doubling the dose to 120 mg/kg/week has been shown to maintain AAT trough levels in the normal range without an increase in side effects [58]. The clinical benefits of this alternate dosing regimen are currently being evaluated in the largest international AATD trial to date, the Study of ProlAstin-C Randomized Therapy with Alpha-1 augmentation (SPARTA) [59].

Aerosolized AAT has gained attention as it may provide direct administration of AAT in the target organ at a lower dose and cost and with greater patient acceptability. Small studies have demonstrated that aerosolized AAT delivery leads to significant elevations of AAT above protective levels in lung epithelial lining fluid with a once or twice a day inhalation [60]. In patients with AATD and relatively mild lung function impairment, adequate deposition of A1AT in the lung periphery appears to be achievable [61]. Larger trials testing its clinical efficacy are pending.

Gene-based therapeutics has been studied as an alternative to augment AAT levels. With the use of an adeno-associated virus (AAV), the normal AAT gene has been successfully introduced into striated muscle in humans and proven to be safe and with sustained (but low) AAT levels at 12 months [62]. Intrapleural delivery of AAV vectors in murine and primate model systems have shown long-term pleural expression of AAT mRNA up to 1 year posttreatment [63]. The normal AAT gene has also been delivered directly to the respiratory epithelium (nostril) encased in a plasmid–cationic liposome complex and shown to produce potentially local therapeutic AAT concentrations and anti-inflammatory effects.

2   Conclusions

Current evidence based on prospective observational cohorts and newer randomized studies suggest that intravenous augmentation therapy has a positive impact in slowing lung function decline and slowing emphysema progression. These effects are supported by biochemical evidence of increased anti-elastase protection and a decrease in lung inflammation. Experience with intravenous augmentation therapy suggests that it is safe with few, and usually well-tolerated side effects.

The decision to initiate augmentation therapy in affected individuals with impaired lung function should be based on the collective data of efficacy, safety of augmentation therapy, and the current lack of other alternatives. A personalized approach has been encouraged. Adequately powered randomized, placebo-controlled trials, utilizing clinical and biochemical endpoints of efficacy are needed to determine the most effective augmentation regimen and the subgroups of patients who would benefit from it. Attempts to improve the cost effectiveness of therapy should be made, including alternate modalities of AAT delivery.

References

1. Campos MA, Alazemi S, Zhang G, Wanner A, Sandhaus RA (2009) Effects of a disease management program in individuals with alpha-1 antitrypsin deficiency. COPD 6(1):31–40. doi:10.1080/15412550802607410

2. American Thoracic Society/European Respiratory Society Statement (2003) Standards for the diagnosis and Management of Individuals with alpha-1 antitrypsin deficiency. Am J Respir Crit Care Med 168(7):818–900

3. Hoffmann U, Bergler T, Rihm M, Pace C, Kruger B, Rummele P, Stoelcker B, Banas B, Mannel DN, Kramer BK (2009) Upregulation of TNF receptor type 2 in human and experimental renal allograft rejection. Am J Transplant 9(4):675–686. doi:10.1111/j.1600-6143.2008.02536.x

4. Lutfi RA, Liu CJ, Stoelinga C (2008) Level dominance in sound source identification. J Acoust Soc Am 124(6):3784–3792. doi:10.1121/1.2998767

5. Stoelting S, Trefzer T, Kisro J, Steinke A, Wagner T, Peters SO (2008) Low-dose oral metronomic chemotherapy prevents mobilization of endothelial progenitor cells into the blood of cancer patients. In Vivo 22(6):831–836

6. Lee S, Goh BT, Tideman H, Stoelinga PJ, Jansen JA (2009) Modular endoprosthesis for mandibular body reconstruction: a clinical, micro-CT and histologic evaluation in eight *Macaca fascicularis*. Int J Oral Maxillofac Surg 38(1):40–47. doi:10.1016/j.ijom.2008.11.020

7. Verdonck HW, Meijer GJ, Kessler P, Nieman FH, de Baat C, Stoelinga PJ (2009) Assessment of bone vascularity in the anterior mandible using laser Doppler flowmetry. Clin Oral Implants Res 20(2):140–144. doi:10.1111/j.1600-0501.2008.01631.x

8. Stoelting M, Geyer M, Reuter S, Reichelt R, Bek MJ, Pavenstadt H (2009) Alpha/beta hydrolase 1 is upregulated in D5 dopamine receptor knockout mice and reduces O2- production of NADPH oxidase. Biochem Biophys Res Commun 379(1):81–85. doi:10.1016/j.bbrc.2008.12.008

9. Crystal RG (1990) Alpha 1-antitrypsin deficiency, emphysema, and liver disease. Genetic basis and strategies for therapy. J Clin Invest 85(5):1343–1352. doi:10.1172/JCI114578

10. Wewers MD, Casolaro MA, Sellers SE, Swayze SC, McPhaul KM, Wittes JT, Crystal RG (1987) Replacement therapy for alpha 1-antitrypsin deficiency associated with emphysema. N Engl J Med 316(17):1055–1062

11. Wewers MD, Casolaro MA, Crystal RG (1987) Comparison of alpha-1-antitrypsin levels and antineutrophil elastase capacity of blood and lung in a patient with the alpha-1-antitrypsin phenotype null-null before and during alpha-1-antitrypsin augmentation therapy. Am Rev Respir Dis 135(3):539–543

12. Ma S, Lin YY, Turino GM (2007) Measurements of desmosine and isodesmosine by mass spectrometry in COPD. Chest 131(5):1363–1371. doi:10.1378/chest.06-2251

13. Fregonese L, Ferrari F, Fumagalli M, Luisetti M, Stolk J, Iadarola P (2011) Long-term variability of desmosine/isodesmosine as biomarker in alpha-1-antitrypsin deficiency-related COPD. COPD 8(5):329–333. doi:10.3109/15412555.2011.589871

14. Ma S, Lin YY, He J, Rouhani FN, Brantly M, Turino GM (2013) Alpha-1 antitrypsin augmentation therapy and biomarkers of elastin degradation. COPD 10(4):473–481. doi:10.3109/15412555.2013.771163

15. Kraaier K, Hartmann M, Stoel MG, von Birgelen C (2008) Intermittent spastic coronary occlusion at site of non-significant atherosclerotic lesion requiring stent implantation. Neth Heart J 16(11):390–391

16. Carter GT, Jensen MP, Hoffman AJ, Stoelb BL, Abresch RT, McDonald CM (2008) Pain in myotonic muscular dystrophy, type 1. Arch Phys Med Rehabil 89(12):2382. doi:10.1016/j.apmr.2008.09.001

17. Jonigk D, Al-Omari M, Maegel L, Muller M, Izykowski N, Hong J, Hong K, Kim SH, Dorsch M, Mahadeva R, Laenger F, Kreipe H, Braun A, Shahaf G, Lewis EC, Welte T, Dinarello CA, Janciauskiene S (2013) Anti-inflammatory and immunomodulatory properties of alpha1-antitrypsin without inhibition of elastase. Proc Natl Acad Sci U S A. doi:10.1073/pnas.1309648110

18. Petrache I, Fijalkowska I, Zhen L, Medler TR, Brown E, Cruz P, Choe KH, Taraseviciene-Stewart L, Scerbavicius R, Shapiro L, Zhang B, Song S, Hicklin D, Voelkel NF, Flotte T, Tuder RM (2006) A novel antiapoptotic role for alpha1-antitrypsin in the prevention of pulmonary emphysema. Am J Respir Crit Care Med 173(11):1222–1228. doi:10.1164/rccm.200512-1842OC

19. Rouhani F, Paone G, Smith NK, Krein P, Barnes P, Brantly ML (2000) Lung neutrophil burden correlates with increased proinflammatory cytokines and decreased lung function in individuals with alpha(1)-antitrypsin deficiency. Chest 117(5 Suppl 1):250S–251S

20. Bergin DA, Reeves EP, Meleady P, Henry M, McElvaney OJ, Carroll TP, Condron C, Chotirmall SH, Clynes M, O'Neill SJ, McElvaney NG (2010) Alpha-1 antitrypsin regulates human neutrophil chemotaxis induced by soluble immune complexes and IL-8. J Clin Invest 120(12):4236–4250. doi:10.1172/JCI41196

21. Bergin DA, Reeves EP, Hurley K, Wolfe R, Jameel R, Fitzgerald S, McElvaney NG (2014) The circulating proteinase inhibitor alpha-1 antitrypsin regulates neutrophil degranulation and autoimmunity. Sci Transl Med 6(217):217ra211. doi:10.1126/scitranslmed.3007116

22. Schmid ST, Koepke J, Dresel M, Hattesohl A, Frenzel E, Perez J, Lomas DA, Miranda E, Greulich T, Noeske S, Wencker M, Teschler H, Vogelmeier C, Janciauskiene S, Koczulla AR (2012) The effects of weekly augmentation therapy in patients with PiZZ alpha1-antitrypsin deficiency. Int J Chron Obstruct Pulmon Dis 7:687–696. doi:10.2147/COPD.S34560

23. Stockley RA, Bayley DL, Unsal I, Dowson LJ (2002) The effect of augmentation therapy on bronchial inflammation in alpha1-antitrypsin deficiency. Am J Respir Crit Care Med 165(11):1494–1498. doi:10.1164/rccm.2109013

24. Seersholm N, Wencker M, Banik N, Viskum K, Dirksen A, Kok-Jensen A, Konietzko N (1997) Does alpha1-antitrypsin augmentation therapy slow the annual decline in FEV1 in patients with severe hereditary alpha1-antitrypsin deficiency? Wissenschaftliche Arbeitsgemeinschaft zur Therapie von Lungenerkrankungen (WATL) alpha1-AT study group. Eur Respir J 10(10):2260–2263

25. Survival and FEV1 decline in individuals with severe deficiency of alpha1-antitrypsin (1998) The alpha-1-antitrypsin deficiency registry study group. Am J Respir Crit Care Med 158(1):49–59

26. Wencker M, Fuhrmann B, Banik N, Konietzko N (2001) Longitudinal follow-up of patients with alpha(1)-protease inhibitor deficiency before and during therapy with IV alpha(1)-protease inhibitor. Chest 119(3):737–744

27. Chapman KR, Bradi AC, Paterson D, Navickis RJ, Wilkes MM (2005) Slower lung function decline during augmentation therapy in patients with alpha-1 antitrypsin deficiency: results from the Canadian AIR registry. Proc Am Thorac Soc 2:A808

28. Chapman KR, Stockley RA, Dawkins C, Wilkes MM, Navickis RJ (2009) Augmentation therapy for alpha1 antitrypsin deficiency: a meta-analysis. COPD 6(3):177–184

29. Schluchter MD, Stoller JK, Barker AF, Buist AS, Crystal RG, Donohue JF, Fallat RJ, Turino GM, Vreim CE, Wu MC (2000) Feasibility of a clinical trial of augmentation therapy for alpha (1)-antitrypsin deficiency. The alpha 1-antitrypsin deficiency registry study group. Am J Respir Crit Care Med 161(3 Pt 1):796–801. doi:10.1164/ajrccm.161.3.9906011

30. Dirksen A, Friis M, Olesen KP, Skovgaard LT, Sorensen K (1997) Progress of emphysema in severe alpha 1-antitrypsin deficiency as assessed by annual CT. Acta Radiol 38(5):826–832

31. Dowson LJ, Newall C, Guest PJ, Hill SL, Stockley RA (2001) Exercise capacity predicts health status in alpha(1)-antitrypsin deficiency. Am J Respir Crit Care Med 163(4):936–941

32. Dawkins PA, Dowson LJ, Guest PJ, Stockley RA (2003) Predictors of mortality in alpha1-antitrypsin deficiency. Thorax 58(12):1020–1026

33. Dirksen A, Dijkman JH, Madsen F, Stoel B, Hutchison DC, Ulrik CS, Skovgaard LT, Kok-Jensen A, Rudolphus A, Seersholm N, Vrooman HA, Reiber JH, Hansen NC, Heckscher T, Viskum K, Stolk J (1999) A randomized clinical trial of alpha(1)-antitrypsin augmentation therapy. Am J Respir Crit Care Med 160(5 Pt 1):1468–1472

34. Dirksen A, Piitulainen E, Parr DG, Deng C, Wencker M, Shaker SB, Stockley RA (2009) Exploring the role of CT densitometry: a randomised study of augmentation therapy in alpha1-antitrypsin deficiency. Eur Respir J 33(6):1345–1353. doi:10.1183/09031936.00159408

35. Stockley RA, Parr DG, Piitulainen E, Stolk J, Stoel BC, Dirksen A (2010) Therapeutic efficacy of alpha-1 antitrypsin augmentation therapy on the loss of lung tissue: an integrated analysis of 2 randomised clinical trials using computed tomography densitometry. Respir Res 11:136. doi:10.1186/1465-9921-11-136

36. Gotzsche PC, Johansen HK (2010) Intravenous alpha-1 antitrypsin augmentation therapy for treating patients with alpha-1 antitrypsin deficiency and lung disease. Cochrane Database Syst Rev 7:CD007851. doi:10.1002/14651858.CD007851.pub2

37. Chapman KR, Burdon JG, Piitulainen E, Sandhaus RA, Seersholm N, Stocks JM, Stoel BC, Huang L, Yao Z, Edelman JM, McElvaney NG, RAPID Trial Study Group (2015) Intravenous augmentation treatment and lung density in severe alpha1 antitrypsin deficiency (RAPID): a randomised, double-blind, placebo-controlled trial. Lancet 386(9991):360–368. doi:10.1016/S0140-6736(15)60860-1

38. McElvaney NG, Burdon J, Holmes M, Glanville A, Wark PA, Thompson PJ, Hernandez P, Chlumsky J, Teschler H, Ficker JH, Seersholm N, Altraja A, Mäkitaro R, Chorostowska-Wynimko J, Sanak M, Stoicescu PI, Piitulainen E, Vit O, Wencker M, Tortorici MA, Fries M, Edelman JM, Chapman KR, RAPID Extension Trial Group (2017) Long-term efficacy and safety of α1 proteinase inhibitor treatment for emphysema caused by severe α1 antitrypsin deficiency: an open-label extension trial (RAPID-OLE). Lancet Respir Med 5 (1):51–60. doi:10.1016/S2213-2600(16) 30430-1

39. Lieberman J (2000) Augmentation therapy reduces frequency of lung infections in antitrypsin deficiency: a new hypothesis with supporting data. Chest 118(5):1480–1485

40. Campos MA, Sandhaus R (2004) Exacerbations of respiratory symptoms in patients with alpha-1 antitrypsin deficiency on augmentation therapy. Am J Respir Crit Care Med 169(7): A767

41. Needham M, Stockley RA (2005) Exacerbations in {alpha}1-antitrypsin deficiency. Eur Respir J 25(6):992–1000. doi:10.1183/ 09031936.05.00074704

42. Wencker M, Banik N, Buhl R, Seidel R, Konietzko N (1998) Long-term treatment of alpha1-antitrypsin deficiency-related pulmonary emphysema with human alpha1-antitrypsin. Wissenschaftliche Arbeitsgemeinschaft zur Therapie von Lungenerkrankungen (WATL)-alpha1-AT-study group. Eur Respir J 11(2):428–433

43. Stoller JK, Fallat R, Schluchter MD, O'Brien RG, Connor JT, Gross N, O'Neil K, Sandhaus R, Crystal RG (2003) Augmentation therapy with alpha1-antitrypsin: patterns of use and adverse events. Chest 123(5):1425–1434

44. Meyer FJ, Wencker M, Teschler H, Steveling H, Sennekamp J, Costabel U, Konietzko N (1998) Acute allergic reaction and demonstration of specific IgE antibodies against alpha-1-protease inhibitor. Eur Respir J 12(4):996–997

45. Mullins CD, Huang X, Merchant S, Stoller JK, Alpha One Foundation Research Network Registry I (2001) The direct medical costs of alpha(1)-antitrypsin deficiency. Chest 119 (3):745–752

46. Mullins CD, Wang J, Stoller JK (2003) Major components of the direct medical costs of alpha1-antitrypsin deficiency. Chest 124 (3):826–831

47. Mullins CD, Blatt L, Wang J (2002) Societal implications of the pharmacoeconomics of alpha1-antitrypsin deficiency. Expert Rev Pharmacoecon Outcomes Res 2(3):243–249. doi:10.1586/14737167.2.3.243

48. Sclar DA, Evans MA, Robison LM, Skaer TL (2012) alpha1-proteinase inhibitor (human) in the treatment of hereditary emphysema secondary to alpha1-antitrypsin deficiency: number and costs of years of life gained. Clin Drug Investig 32(5):353–360. doi:10.2165/ 11631920-000000000-00000

49. Alkins SA, O'Malley P (2000) Should healthcare systems pay for replacement therapy in patients with alpha(1)-antitrypsin deficiency? A critical review and cost-effectiveness analysis. Chest 117(3):875–880

50. Gildea TR, Shermock KM, Singer ME, Stoller JK (2003) Cost-effectiveness analysis of augmentation therapy for severe alpha1-antitrypsin deficiency. Am J Respir Crit Care Med 167(10):1387–1392. doi:10.1164/ rccm.200209-1035OC

51. Stockley RA, Miravitlles M, Vogelmeier C (2013) Augmentation therapy for alpha-1 antitrypsin deficiency: towards a personalised approach. Orphanet J Rare Dis 8(1):149. doi:10.1186/1750-1172-8-149

52. Sandhaus RA, Turino G, Stocks J, Strange C, Trapnell BC, Silverman EK, Everett SE, Stoller JK (2008) Alpha1-Antitrypsin augmentation therapy for PI*MZ heterozygotes: a cautionary note. Chest 134(4):831–834. doi:10.1378/ chest.08–0868

53. Campos MA, Alazemi S, Zhang G, Wanner A, Salathe M, Baier H, Sandhaus RA (2009) Exacerbations in subjects with alpha-1 antitrypsin deficiency receiving augmentation therapy. Respir Med 103(10):1532–1539. doi:10. 1016/j.rmed.2009.04.008

54. Hubbard RC, Sellers S, Czerski D, Stephens L, Crystal RG (1988) Biochemical efficacy and safety of monthly augmentation therapy for alpha 1-antitrypsin deficiency. JAMA 260 (9):1259–1264

55. Barker AF, Iwata-Morgan I, Oveson L, Roussel R (1997) Pharmacokinetic study of alpha1-antitrypsin infusion in alpha1-antitrypsin deficiency. Chest 112(3):607–613

56. Soy D, de la Roza C, Lara B, Esquinas C, Torres A, Miravitlles M (2006) Alpha-1-antitrypsin deficiency: optimal therapeutic regimen based on population pharmacokinetics. Thorax 61(12):1059–1064. doi:10.1136/thx.2005. 057943

57. Zamora NP, Pla RV, Del Rio PG, Margaleff RJ, Frias FR, Ronsano JB (2008) Intravenous human plasma-derived augmentation therapy in alpha 1-antitrypsin deficiency: from pharmacokinetic analysis to individualizing therapy.

Ann Pharmacother 42(5):640–646. doi:10. 1345/aph.1K505

58. Campos MA, Kueppers F, Stocks JM, Strange C, Chen J, Griffin R, Wang-Smith L, Brantly ML (2013) Safety and pharmacokinetics of 120 mg/kg versus 60 mg/kg weekly intravenous infusions of alpha-1 proteinase inhibitor in alpha-1 antitrypsin deficiency: a multicenter, randomized, double-blind, crossover study (SPARK). COPD 10(6):687–695. doi:10. 3109/15412555.2013.800852

59. Stoel BC, Bode F, Rames A, Soliman S, Reiber JH, Stolk J (2008) Quality control in longitudinal studies with computed tomographic densitometry of the lungs. Proc Am Thorac Soc 5 (9):929–933.   doi:10.1513/pats.200804-039QC

60. Hubbard RC, Crystal RG (1990) Strategies for aerosol therapy of alpha 1-antitrypsin deficiency by the aerosol route. Lung 168(Suppl 1): 565–578

61. Vogelmeier C, Kirlath I, Warrington S, Banik N, Ulbrich E, Du Bois RM (1997) The intrapulmonary half-life and safety of aerosolized alpha1-protease inhibitor in normal volunteers. Am J Respir Crit Care Med 155 (2):536–541

62. Brantly ML, Spencer LT, Humphries M, Conlon TJ, Spencer CT, Poirier A, Garlington W, Baker D, Song S, Berns KI, Muzyczka N, Snyder RO, Byrne BJ, Flotte TR (2006) Phase I trial of intramuscular injection of a recombinant adeno-associated virus serotype 2 alpha1-antitrypsin (AAT) vector in AAT-deficient adults. Hum Gene Ther 17(12):1177–1186. doi:10.1089/hum.2006.17.1177

63. Chiuchiolo MJ, Kaminsky SM, Sondhi D, Hackett NR, Rosenberg JB, Frenk EZ, Hwang Y, Van de Graaf BG, Hutt JA, Wang G, Benson J, Crystal RG (2013) Intrapleural administration of an AAVrh.10 vector coding for human alpha1-antitrypsin for the treatment of alpha1-antitrypsin deficiency. Hum Gene Ther Clinical Dev 24(4):161–173. doi:10. 1089/humc.2013.168

Chapter 26

Therapeutic Options in Alpha-1 Antitrypsin Deficiency: Liver Transplantation

Nedim Hadzic

Abstract

Alpha-1 antitrypsin deficiency is the commonest genetic condition leading to liver transplantation in childhood. It remains unclear why only a minority of individuals carrying homozygous PiZ phenotype has liver disease, but also why of those only about a quarter develops end stage liver disease, requiring liver transplantation. This intervention has now become routine worldwide with 1-year patient survival rates well above 90%. As for all autosomal recessive conditions liver donation from anonymous cadaveric sources is preferred to living related parental donors, due to their presumed heterozygous state.

Key words Alpha-1 antitrypsin deficiency, Liver transplantation, Immune suppression

1 Introduction

While the quest to find medical means of modifying natural history of alpha-1 antitrypsin deficiency (A1ATD) continues, at present the only effective intervention for individuals developing end-stage liver disease is liver transplantation (LT). Over the last 30 years or so this procedure has firmly established itself with the reported 1-year survival rates ranging between 85 and 95% in most of the transplant centers worldwide [1–5]. In USA, United Network for Organ Sharing (UNOS) reported that A1ATD represented indication for LT in around 1% of adult and 3.5% of pediatric recipients [4]. Some studies noted male prevalence amongst liver transplant recipients [1, 4].

In pediatric medicine the majority of children requiring LT will do so within the first 2–3 years of life. It still remains unclear why only 10–15% of children with homozygous PiZ phenotype, potentially identifiable in the community, develop symptomatic liver disease [6], and also why only about a quarter of those goes on to require LT [7]. There have been some reports suggesting that the A1ATD children presenting with severe and more prolonged neonatal jaundice and more advanced histological changes, such as

Florie Borel and Christian Mueller (eds.), *Alpha-1 Antitrypsin Deficiency: Methods and Protocols*, Methods in Molecular Biology, vol. 1639, DOI 10.1007/978-1-4939-7163-3_26, © Springer Science+Business Media LLC 2017

early fibrosis and bile duct reduplication, could be more likely candidates for the liver replacement [1]. It is noteworthy that in infancy the life-threatening decompensation of chronic liver disease secondary to A1ATD can be fairly rapid following an initial episode of ascites or gastrointestinal bleeding. This is why these children need earlier work up for LT in comparison to other pediatric indications, such as biliary atresia.

Chronic shortage of cadaveric donors frequently leads to a consideration for using living-related donors, usually the patient's parents. However, for many autosomal recessive inherited conditions, including A1ATD, this option remains debatable, due to the perceived long-term risks of possessing a carrier state. For the LT team balancing these risks can be challenging, particularly in light of a child with a rapidly deteriorating clinical condition.

Most children would receive a partial liver graft with careful anastomoses of portal and hepatic vessels and biliary reconnection with Roux-en-Y loop. The explanted liver typically shows features of advanced cirrhosis, with abundant alpha-1 globules throughout the liver lobule. Frequent post-LT complications include infections, cellular rejection, and biliary complications leading to chronic bile leakage, requiring interventional radiology or surgical correction. Post-LT management of individuals with A1ATD, in addition to standard problems, can be complicated by considerable surgical drain losses due to underlying portal hypertension and systemic hypertension potentiated by early use of steroids. The graft rejection is treated by standard agents, if required. The pulmonary complications are not commonplace, as many children reach LT with no respiratory involvement whatsoever. For elective LT indications such as A1ATD, the average length of postoperative stay in hospital is between 2 and 3 weeks.

The alpha-1 antitrypsin phenotype of the donor is established once normal liver graft function is achieved, usually within a few weeks after the operation, resulting in the normal serum levels. This would theoretically indicate that the risks for developing lung disease have been removed. While the follow-up of liver transplanted children is still relatively short to confirm that, one large study from adult A1ATD patients reported that the forced expiratory volume (FEV)-1 continues to decline despite a good liver graft function [5]. It is noteworthy that the development of end-stage lung and liver disease, requiring dual organ transplantation, is extremely rare [8], indirectly supporting the involvement of different pathophysiologic mechanisms.

As A1ATD is increasingly recognized as a cause of end-stage liver disease, including hepatocellular carcinoma, more questions about pathogenic role of a heterozygous PiZ state in adulthood are posed. This is particularly unclear in the setting where this is associated with other potential confounding etiologies such as nonalcoholic fatty liver disease, chronic viral infection, diabetes mellitus or alcohol abuse, all becoming more relevant outside pediatric age. One recent study in adults reported that 8% of homozygous PiZ and 90% of PiMZ patients had added pathologies [5]. Overall, around 18% had hepatocellular carcinoma as the main indication for LT [5]. This area definitely needs further epidemiological observations and more basic scientific research to improve our understanding of these links.

Post-LT management requires a prolonged, often indefinite immunosuppression, with tacrolimus used either as a monotherapy or a backbone of other regimens. This effective medication carries considerable dose-dependent nephrotoxicity risks, potentially jeopardizing the long-term success of LT. Therefore, in the setting of A1ATD, where subclinical renal impairment may be quiescent for many years, renal-sparing regimens, including additional medications combined with lower doses of tacrolimus, are often indicated. Regular post-LT monitoring of renal function is strongly recommended.

LT is undoubtedly an effective intervention for a small proportion of affected homozygous PiZ individuals. However, a decision to pursue this option should always be balanced against the risks of long-term immunosuppression and the increased prevalence of infection and malignancies.

References

1. Francavilla R, Castellaneta SP, Hadzic N, Chambers SM, Portmann B, Tung J, Cheeseman P, Rela M, Heaton ND, Mieli-Vergani G (2000) Prognosis of alpha-1-antitrypsin deficiency-related liver disease in the era of paediatric liver transplantation. J Hepatol 32(6):986–992

2. Vennarecci G, Gunson BK, Ismail T, Hübscher SG, Kelly DA, McMaster P, Elias E (1996) Transplantation for end stage liver disease related to alpha 1 antitrypsin. Transplantation 61(10):1488–1495

3. Filipponi F, Soubrane O, Labrousse F, Devictor D, Bernard O, Valayer J, Houssin D (1994) Liver transplantation for end-stage liver disease associated with alpha-1-antitrypsin deficiency in children: pretransplant natural history, timing and results of transplantation. J Hepatol 20 (1):72–78

4. Kemmer N, Kaiser T, Zacharias V, Neff GW (2008) Alpha-1-antitrypsin deficiency: outcomes after liver transplantation. Transplant Proc 40(5):1492–1494

5. Carey EJ, Iyer VN, Nelson DR, Nguyen JH, Krowka MJ (2013) Outcomes for recipients of liver transplantation for alpha-1-antitrypsin deficiency–related cirrhosis. Liver Transpl 19 (12):1370–1376

6. Sveger T (1976) Liver disease in alpha1-antitrypsin deficiency detected by screening of 200,000 infants. N Engl J Med 294 (24):1316–1321

7. Hussain M, Mieli-Vergani G, Mowat AP (1991) Alpha 1-antitrypsin deficiency and liver disease: clinical presentation, diagnosis and treatment. J Inherit Metab Dis 14(4):497–511

8. Dawwas MF, Davies SE, Griffiths WJ, Lomas DA, Alexander GJ (2013) Prevalence and risk factors for liver involvement in individuals with PiZZ-related lung disease. Am J Respir Crit Care Med 187(5):502–508

Chapter 27

Therapeutics: Gene Therapy for Alpha-1 Antitrypsin Deficiency

Alisha M. Gruntman and Terence R. Flotte

Abstract

This review seeks to give an overview of alpha-1 antitrypsin deficiency, including the different disease phenotypes that it encompasses. We then describe the different therapeutic endeavors that have been undertaken to address these different phenotypes. Lastly we discuss future potential therapeutics, such as genome editing, and how they may play a role in treating alpha-1 antitrypsin deficiency.

Key words SERPINA1, Emphysema, Liver disease, Genome editing, CRISPR/Cas9

1 Introduction

Alpha-1 antitrypsin (AAT) deficiency has been viewed as an accessible target for gene therapy because the gene product is a secreted serum protein whose level is readily measurable in peripheral blood. In fact, the currently available protein replacement products have used plasma or serum AAT levels as the endpoint for FDA licensure, and it seems likely that any gene therapy for the lung disease in AAT deficiency would do likewise. Furthermore, AAT levels naturally have a wide concentration range, indicating that precise regulation of expression would not be required. Countering these advantages is the fact that AAT is second only to albumin among the most abundant circulating serum proteins in humans. The threshold level for protection from lung disease is 11 μM (approximately 571 mcg/ml or 57 mg/dl). AAT deficiency is remarkably homogeneous genetically, with over 90% of mutant alleles due to one particular missense mutation known as PiZ (Glu342Lys, E342K).

Further complexity derives from the fact that the deficiency state, resulting in chronic lung disease in these patients, is accompanied by liver disease, due to accumulation of the Z-mutant protein, in a subset of approximately 10% of patients. Gene therapy approaches for AAT liver disease would have a different molecular

Florie Borel and Christian Mueller (eds.), *Alpha-1 Antitrypsin Deficiency: Methods and Protocols*, Methods in Molecular Biology, vol. 1639, DOI 10.1007/978-1-4939-7163-3_27, © Springer Science+Business Media LLC 2017

target and endpoint. The goal of molecular therapy for AAT liver disease would be to reduce the steady-state concentration of Z-AAT within hepatocytes, while the therapy for lung disease would consist of augmentation of the wild type (or PiM) protein from any tissue or cell type capable of AAT secretion. The recent licensure of a rAAV1-lipoprotein lipase vector (Glybera[R]) in Europe has raised expectations for gene therapy, but the choice of vector and cargo for treatment of AAT deficiency will be dictated by the molecular endpoints required to impact the lung disease, the liver disease, or both.

2   Physiologic Functions of Alpha-1 Antitrypsin (AAT)

AAT is the product of the SERPINA1 gene, one of a family of serine proteinase inhibitors (serpins), which perform multiple and diverse regulatory functions. Across species, serpins often function to modulate protease cascades in inflammatory and coagulation pathways. Important serpins in humans include anti-thrombin III (ATIII), which inhibits thrombin, a key enzyme in both the intrinsic and extrinsic coagulation cascades. Interestingly, one naturally occurring mutant form of AAT [Pittsburgh 358 Met → Arg] has anti-thrombin activity and causes a bleeding diathesis in affected patients [1].

AAT is the most abundant serum serpin in humans, with physiologic levels ranging from approximately 11–25 μM. AAT is an acute phase reactant and has an IL-6-responsive promoter, induced by a variety of infectious and inflammatory stimuli, including fever and viral infections. Neutrophil elastase (NE) is the principal substrate for AAT in serum and tissues, but AAT also has a wide range of anti-protease activities, with demonstrated ability to inhibit certain cathepsins, proteinase-3, alpha-defensins, and caspase-3 (*see* Table 1) [2–4].

Table 1
Some putative functions of alpha-1 antitrypsin

Neutrophil elastase inhibition
Cathepsin-G inhibition
Proteinase-3 inhibition
Caspase-3 inhibition (anti-apoptotic)
IL-8 binding
Alpha-defensin inhibition
Lipid raft binding
HIV-envelope glycoprotein binding
Direct antioxidant effect

The anti-NE role of AAT in the lung appears to have both primary and secondary anti-inflammatory and anti-apoptotic benefits. AAT has a direct effect on protection of interstitial elastin the lung. AAT-deficient patients have increased levels of elastin breakdown products in their urine and AAT replacement causes reduction of such fragments. The absence of AAT often produces a classic panacinar pattern of emphysema, indicating widespread loss of alveolar septae and their supporting elastin matrix.

AAT's antiprotease role also has secondary benefits in inhibiting inflammation in both the alveolar and airway compartments of the lung [5]. Elastin fragments have an important pro-inflammatory role, triggering secretion of IL-8, a potent neutrophil chemoattractant [6]. AAT also is a multivalent antioxidant, dampening the role of reactive oxygen species (ROS) in propagating inflammatory pathways in the lung and liver [7, 8]. Patients with AAT deficiency have histologic evidence of airway inflammation and measureable elevations of several pro-inflammatory cytokines in epithelial lining fluid (ELF) [9, 10]. There also appears to be increased adaptive immune activation within the lungs of AAT deficient patients [11]. Clinically, the majority of AAT deficient lung disease patients have evidence of asthma-like symptoms with wheezing and reversible airway obstruction, signs and symptoms typical of inflammatory airway disease.

The protective role of AAT for the lungs is particularly crucial in the face of environmental exposures. Tobacco smoking is the major cofactor in determining the clinical severity of AAT-deficient lung disease, implying that a key physiologic function of AAT relates to protection of the lung from inhaled noxious agents, presumably mediated by the antioxidant and antiprotease functions of AAT [12–14].

3   Current Therapy for AAT Deficiency

Lung disease in AAT deficiency presents on the COPD spectrum, primarily emphysema with some asthmatic characteristics. The typical AAT lung disease patient is diagnosed in their early 50s but often has a history of 10 years of preceding pulmonary complaints. Earlier diagnosis of AAT lung disease has the potential for substantial benefit. The central role of avoidance of cigarette smoking and smoking cessation in AAT lung disease therapy cannot be over-emphasized Supportive therapies for AAT lung disease mirror those used for sporadic COPD, including supplemental oxygen, inhaled corticosteroids, inhaled bronchodilators, lung reduction surgery, and lung transplantation.

Specific therapy with IV AAT protein replacement was first demonstrated in the 1980s and has been commercially available since 1989 [9, 15]. Several protein replacement products are currently available in the US and several European countries. Pooled human plasma products for AAT deficiency available in the US include Aralast NP (Baxter), Prolastin-C (Grifols), Zemaira (CSL Behring), and Glassia (Baxter). Other forms of AAT replacement have been tested in humans, including recombinant and transgenic forms, but these have not been licensed for clinical use.

4   AAT Lung Disease Gene Therapy

Current IV protein replacement therapies have been approved based on replacement of serum or plasma levels (with the benchmark level of 11 μM), so that the monitoring of the therapeutic response for an AAT lung disease gene therapy should be straightforward. The generally accepted endpoint for AAT lung disease gene therapy would likewise be to express a sufficient quantity of AAT to establish a steady-state level of at least 11 μM of AAT. While the measurement of this endpoint is straightforward, this is an ambitious target given that the native level of AAT is second only to albumin in abundance among serum proteins.

Since AAT is secreted efficiently from many different cell types, AAT augmentation for AAT gene therapy could potentially be accomplished by delivery of the M-AAT gene to any tissue or organ capable of secreting proteins into the serum. AAT is normally produced within the liver. However, approaches to AAT lung disease gene therapy have included AAT gene delivery to muscle, pleura, lung, and nose, in addition to liver-directed delivery.

In the early 1990s, gene delivery to the lung was proven possible, and subsequently a number of potential gene therapy vectors have been used for AAT gene therapy [16]. Cationic liposome delivery of AAT to several different organs has been demonstrated [17–21] and nasal delivery of liposomal AAT plasmid was evaluated by Brigham et al. [22]. In these studies expression was low and transient. Recombinant adenovirus (rAd) AAT was evaluated preclinically for lung, liver, and muscle delivery [23, 24], resulting in transiently high levels of expression. Substantially more prolonged expression was observed with helper-dependent Ad [25].

5   rAAV-AAT Gene Therapy

Recombinant adeno-associated virus (rAAV) gene therapy for AAT lung disease has also been examined preclinically for delivery to muscle, lung, and liver [26–31]. In addition, pleural-based delivery of AAT has been examined using a number of different serotypes and species [32–34].

Direct intramuscular (IM) injection of rAAV-AAT has been developed for clinical gene therapy of AAT lung disease. Initially, rAAV2-AAT was shown to mediate stable, high-level AAT gene expression in C57Bl6 mice after IM delivery [26] and this was reproduced in nonhuman primates [28]. This supported the initiation of a phase 1 trial of rAAV2-AAT IM delivery in AAT deficient adults. Gene expression was low and transient in that trial [35].

Direct comparison of rAAV2-AAT and rAAV1-AAT in C57Bl6 mice indicated the likelihood that the same expression cassette would mediate a higher level of AAT augmentation when packaged in a rAAV1 capsid. Subsequent toxicology and biodistribution studies supported the initiation of clinical trials of rAAV1-AAT IM [36–38].

The initial phase 1 trial of rAAV1-AAT by IM delivery resulted in stable expression approximately 2000-fold below the therapeutic threshold. Expression was dose-related and stable despite a concomitant AAV capsid-specific effector T-cell response as detected by gamma-interferon ELISPOT assays on peripheral blood [36].

Vector production was changed to a herpes-simplex virus (HSV)-based helper system to allow for more scalable production of higher doses of rAAV1-AAT. The vector produced with this method also demonstrated a higher potency in both preclinical and clinical studies. In the clinical study, vector expression peaked at 30 days at approximately 20-fold below therapeutic (i.e., 5% of target) and then receded to approximately 3% of target and plateaued at that level for 1 year (*see* Fig. 1) [39]. This stable expression was achieved in spite of local cellular infiltration within the injected muscle. The mechanism for this persistence appears to relate to both locally and systemically detected regulatory T cells (T_{regs}) with AAV1 capsid specificity.

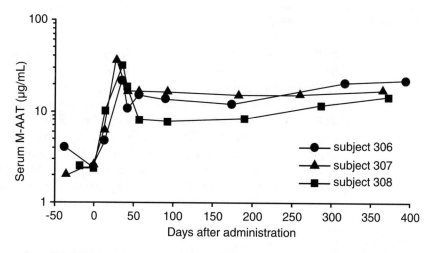

Fig. 1 Expression of M-AAT last over 1 year in AAT-deficient patients after IM administration of a rAAV1-AAT (Reproduced with permission from American Society for Clinical Investigation.)

6 Gene Therapy for AAT Liver Disease

As mentioned above, the therapeutic goal for AAT liver disease is the reduction of endogenous Z-AAT expression. This has been accomplished experimentally in vitro and in Z-AAT transgenic mice using a number of different techniques. Recombinant SV-40 expressing an anti-AAT ribozyme demonstrated robust reduction in the Z-transgenic model [40]. Subsequently, rAAV vectors have been developed expressing anti-AAT short-hairpin RNAs [41] and anti-AAT synthetic miRNAs [42]. In the latter study, the expression of the anti-AAT miRNA was combined with an M-AAT augmentation allele resistant to targeting by that miRNA sequence, resulting in simultaneous allele-specific augmentation and knockdown.

7 Prospects for Genome Editing of AAT Deficiency

The advent of efficient genome editing technology based on zinc-finger nucleases (ZFN), TALENs, and particularly the RNA-guided CRISPR/Cas9 system has allowed for a more practical consideration for definitive gene correction of the Z-AAT mutation in hepatocytes. CRISPR/Cas9 has been shown to be a very high efficiency system for induction of double-strand breaks (DSBs). Targeting to the endogenous AAT locus in the liver would be predicted to create indels which would inactivate the production of Z-AAT in PI*ZZ homozygotes. The induction of DSBs in the presence of donor DNA substrate for homology dependent recombination (an HDR substrate) could be used to effect gene repair, if the HDR template were to span the mutation site.

8 Summary and Future Prospects

AAT has long been seen as a promising target for gene therapy, because it is a relatively common single gene disorder and the endpoint for replacement of serum levels and subsequent lung disease therapy is very straightforward (a serum level of 11 µM). The primary challenges are the very high level of AAT protein that is required and the fact that liver disease therapy will likely require concurrent knockdown of mutant AAT protein. Innovations such as the advent of synthetic miRNA technology and high-efficiency gene editing with CRISPR/Cas9 offer great promise in addressing these challenges.

References

1. Owen MC, Brennan SO, Lewis JH, Carrell RW (1983) Mutation of antitrypsin to antithrombin. N Engl J Med 309(12):694–698. doi:10.1056/NEJM198309223091203

2. Lockett AD, Van Demark M, Gu Y, Schweitzer KS, Sigua N, Kamocki K, Fijalkowska I, Garrison J, Fisher AJ, Serban K, Wise RA, Flotte TR, Mueller C, Presson RG Jr, Petrache HI, Tuder RM, Petrache I (2012) Effect of cigarette smoke exposure and structural modifications on the alpha-1 Antitrypsin interaction with caspases. Mol Med 18:445–454. doi:10.2119/molmed.2011.00207

3. Petrache I, Fijalkowska I, Medler TR, Skirball J, Cruz P, Zhen L, Petrache HI, Flotte TR, Tuder RM (2006) Alpha-1 antitrypsin inhibits caspase-3 activity, preventing lung endothelial cell apoptosis. Am J Pathol 169(4):1155–1166

4. Rouhani F, Paone G, Smith NK, Krein P, Barnes P, Brantly ML (2000) Lung neutrophil burden correlates with increased pro-inflammatory cytokines and decreased lung function in individuals with alpha(1)-antitrypsin deficiency. Chest 117(5 Suppl 1):250s–251s

5. Jonigk D, Al-Omari M, Maegel L, Muller M, Izykowski N, Hong J, Hong K, Kim SH, Dorsch M, Mahadeva R, Laenger F, Kreipe H, Braun A, Shahaf G, Lewis EC, Welte T, Dinarello CA, Janciauskiene S (2013) Anti-inflammatory and immunomodulatory properties of alpha1-antitrypsin without inhibition of elastase. Proc Natl Acad Sci U S A 110 (37):15007–15012. doi:10.1073/pnas.1309648110

6. Aggarwal Y (2006) Elastin fragments are pro-inflammatory in the progression of emphysema. Thorax 61(7):567–567. doi:10.1136/thx.2006.la0205

7. Marcus NY, Blomenkamp K, Ahmad M, Teckman JH (2012) Oxidative stress contributes to liver damage in a murine model of alpha-1-antitrypsin deficiency. Exp Biol Med (Maywood) 237(10):1163–1172. doi:10.1258/ebm.2012.012106

8. Escribano A, Amor M, Pastor S, Castillo S, Sanz F, Codoner-Franch P, Dasi F (2015) Decreased glutathione and low catalase activity contribute to oxidative stress in children with alpha-1 antitrypsin deficiency. Thorax 70 (1):82–83. doi:10.1136/thoraxjnl-2014-205898

9. Gadek JE, Klein HG, Holland PV, Crystal RG (1981) Replacement therapy of alpha-1-antitrypsin deficiency. Reversal of protease-antiprotease imbalance within the alveolar structures of PiZ subjects. J Clin Invest 68 (5):1158–1165

10. Hubbard RC, Fells G, Gadek J, Pacholok S, Humes J, Crystal RG (1991) Neutrophil accumulation in the lung in alpha 1-antitrypsin deficiency. Spontaneous release of leukotriene B4 by alveolar macrophages. J Clin Invest 88 (3):891–897. doi:10.1172/jci115391

11. Baraldo S, Turato G, Lunardi F, Bazzan E, Schiavon M, Ferrarotti I, Molena B, Cazzuffi R, Damin M, Balestro E, Luisetti M, Rea F, Calabrese F, Cosio MG, Saetta M (2015) Immune activation in alpha1-antitrypsin-deficiency emphysema. Beyond the protease-antiprotease paradigm. Am J Respir Crit Care Med 191(4):402–409. doi:10.1164/rccm.201403-0529OC

12. O'Brien ME, Pennycooke K, Carroll TP, Shum J, Fee LT, O'Connor C, Logan PM, Reeves EP, McElvaney NG (2015) The impact of smoke exposure on the clinical phenotype of alpha-1 antitrypsin deficiency in Ireland: exploiting a national registry to understand a rare disease. COPD 12(Suppl 1):2–9. doi:10.3109/15412555.2015.1021913

13. Carp H, Miller F, Hoidal JR, Janoff A (1982) Potential mechanism of emphysema: alpha 1-proteinase inhibitor recovered from lungs of cigarette smokers contains oxidized methionine and has decreased elastase inhibitory capacity. Proc Natl Acad Sci U S A 79 (6):2041–2045

14. Janus ED, Phillips NT, Carrell RW (1985) Smoking, lung function, and alpha 1-antitrypsin deficiency. Lancet (London, England) 1(8421):152–154

15. Wewers MD, Casolaro MA, Sellers SE, Swayze SC, McPhaul KM, Wittes JT, Crystal RG (1987) Replacement therapy for alpha 1-antitrypsin deficiency associated with emphysema. N Engl J Med 316(17):1055–1062. doi:10.1056/nejm198704233161704

16. Yoshimura K, Rosenfeld MA, Nakamura H, Scherer EM, Pavirani A, Lecocq JP, Crystal RG (1992) Expression of the human cystic fibrosis transmembrane conductance regulator gene in the mouse lung after in vivo intratracheal plasmid-mediated gene transfer. Nucleic Acids Res 20(12):3233–3240

17. Alino SF, Crespo J, Bobadilla M, Lejarreta M, Blaya C, Crespo A (1994) Expression of human alpha 1-antitrypsin in mouse after in vivo gene transfer to hepatocytes by small liposomes. Biochem Biophys Res Commun 204(3):1023–1030. doi:10.1006/bbrc.1994.2565

18. Alino SF, Bobadilla M, Crespo J, Lejarreta M (1996) Human alpha 1-antitrypsin gene transfer to in vivo mouse hepatocytes. Hum Gene Ther 7(4):531–536. doi:10.1089/hum.1996. 7.4-531

19. Crepso J, Blaya C, Crespo A, Alino SF (1996) Long-term expression of the human alpha1-antitrypsin gene in mice employing anionic and cationic liposome vectors. Biochem Pharmacol 51(10):1309–1314

20. Alino SF (1997) Long-term expression of the human alpha1-antitrypsin gene in mice employing anionic and cationic liposome vector. Biochem Pharmacol 54(1):9–13

21. Canonico AE, Conary JT, Meyrick BO, Brigham KL (1994) Aerosol and intravenous transfection of human alpha 1-antitrypsin gene to lungs of rabbits. Am J Respir Cell Mol Biol 10 (1):24–29.   doi:10.1165/ajrcmb.10.1. 8292378

22. Brigham KL, Lane KB, Meyrick B, Stecenko AA, Strack S, Cannon DR, Caudill M, Canonico AE (2000) Transfection of nasal mucosa with a normal alpha1-antitrypsin gene in alpha1-antitrypsin-deficient subjects: comparison with protein therapy. Hum Gene Ther 11 (7):1023–1032.   doi:10.1089/ 10430340050015338

23. Mastrangeli A, Danel C, Rosenfeld MA, Stratford-Perricaudet L, Perricaudet M, Pavirani A, Lecocq JP, Crystal RG (1993) Diversity of airway epithelial cell targets for in vivo recombinant adenovirus-mediated gene transfer. J Clin Invest 91(1):225–234. doi:10. 1172/jci116175

24. Siegfried W, Rosenfeld M, Stier L, Stratford-Perricaudet L, Perricaudet M, Pavirani A, Lecocq JP, Crystal RG (1995) Polarity of secretion of alpha 1-antitrypsin by human respiratory epithelial cells after adenoviral transfer of a human alpha 1-antitrypsin cDNA. Am J Respir Cell Mol Biol 12(4):379–384. doi:10.1165/ ajrcmb.12.4.7695917

25. Cerullo V, McCormack W, Seiler M, Mane V, Cela R, Clarke C, Rodgers JR, Lee B (2007) Antigen-specific tolerance of human alpha1-antitrypsin induced by helper-dependent adenovirus. Hum Gene Ther 18(12):1215–1224. doi:10.1089/hum.2006.036

26. Song S, Morgan M, Ellis T, Poirier A, Chesnut K, Wang J, Brantly M, Muzyczka N, Byrne BJ, Atkinson M, Flotte TR (1998) Sustained secretion of human alpha-1-antitrypsin from murine muscle transduced with adeno-associated virus vectors. Proc Natl Acad Sci U S A 95 (24):14384–14388

27. Song S, Embury J, Laipis PJ, Berns KI, Crawford JM, Flotte TR (2001) Stable therapeutic serum levels of human alpha-1 antitrypsin (AAT) after portal vein injection of recombinant adeno-associated virus (rAAV) vectors. Gene Ther 8(17):1299–1306. doi:10.1038/ sj.gt.3301422

28. Song S, Scott-Jorgensen M, Wang J, Poirier A, Crawford J, Campbell-Thompson M, Flotte TR (2002) Intramuscular administration of recombinant adeno-associated virus 2 alpha-1 antitrypsin (rAAV-SERPINA1) vectors in a nonhuman primate model: safety and immunologic aspects. Mol Ther 6(3):329–335

29. Lu Y, Choi YK, Campbell-Thompson M, Li C, Tang Q, Crawford JM, Flotte TR, Song S (2006) Therapeutic level of functional human alpha 1 antitrypsin (hAAT) secreted from murine muscle transduced by adeno-associated virus (rAAV1) vector. J Gene Med 8 (6):730–735. doi:10.1002/jgm.896

30. Liqun Wang R, McLaughlin T, Cossette T, Tang Q, Foust K, Campbell-Thompson M, Martino A, Cruz P, Loiler S, Mueller C, Flotte TR (2009) Recombinant AAV serotype and capsid mutant comparison for pulmonary gene transfer of alpha-1-antitrypsin using invasive and noninvasive delivery. Mol Ther 17 (1):81–87. doi:10.1038/mt.2008.217

31. Conlon TJ, Cossette T, Erger K, Choi YK, Clarke T, Scott-Jorgensen M, Song S, Campbell-Thompson M, Crawford J, Flotte TR (2005) Efficient hepatic delivery and expression from a recombinant adeno-associated virus 8 pseudotyped alpha1-antitrypsin vector. Mol Ther 12(5):867–875. doi:10.1016/j.ymthe.2005.05.016

32. De BP, Heguy A, Hackett NR, Ferris B, Leopold PL, Lee J, Pierre L, Gao G, Wilson JM, Crystal RG (2006) High levels of persistent expression of alpha1-antitrypsin mediated by the nonhuman primate serotype rh.10 adeno-associated virus despite preexisting immunity to common human adeno-associated viruses. Mol Ther 13(1):67–76. doi:10.1016/j. ymthe.2005.09.003

33. Chiuchiolo MJ, Kaminsky SM, Sondhi D, Hackett NR, Rosenberg JB, Frenk EZ, Hwang Y, Van de Graaf BG, Hutt JA, Wang G, Benson J, Crystal RG (2013) Intrapleural administration of an AAVrh.10 vector coding for human alpha1-antitrypsin for the treatment of alpha1-antitrypsin deficiency. Hum Gene Ther Clin Dev 24(4):161–173. doi:10.1089/ humc.2013.168

34. De B, Heguy A, Leopold PL, Wasif N, Korst RJ, Hackett NR, Crystal RG (2004) Intrapleural administration of a serotype 5 adeno-associated virus coding for alpha1-antitrypsin mediates persistent, high lung and serum levels

of alpha1-antitrypsin. Mol Ther 10 (6):1003–1010. doi:10.1016/j.ymthe.2004. 08.022

35. Brantly ML, Spencer LT, Humphries M, Conlon TJ, Spencer CT, Poirier A, Garlington W, Baker D, Song S, Berns KI, Muzyczka N, Snyder RO, Byrne BJ, Flotte TR (2006) Phase I trial of intramuscular injection of a recombinant adeno-associated virus serotype 2 alpha1-antitrypsin (AAT) vector in AAT-deficient adults. Hum Gene Ther 17(12):1177–1186. doi:10.1089/hum.2006.17.1177

36. Brantly ML, Chulay JD, Wang L, Mueller C, Humphries M, Spencer LT, Rouhani F, Conlon TJ, Calcedo R, Betts MR, Spencer C, Byrne BJ, Wilson JM, Flotte TR (2009) Sustained transgene expression despite T lymphocyte responses in a clinical trial of rAAV1-AAT gene therapy. Proc Natl Acad Sci U S A 106 (38):16363–16368. doi:10.1073/pnas. 0904514106

37. Flotte TR, Conlon TJ, Poirier A, Campbell-Thompson M, Byrne BJ (2007) Preclinical characterization of a recombinant adeno-associated virus type 1-pseudotyped vector demonstrates dose-dependent injection site inflammation and dissemination of vector genomes to distant sites. Hum Gene Ther 18 (3):245–256. doi:10.1089/hum.2006.113

38. Flotte TR, Trapnell BC, Humphries M, Carey B, Calcedo R, Rouhani F, Campbell-Thompson M, Yachnis AT, Sandhaus RA, McElvaney NG, Mueller C, Messina LM, Wilson JM, Brantly M, Knop DR, Ye GJ, Chulay JD (2011) Phase 2 clinical trial of a recombinant adeno-associated viral vector expressing alpha1-antitrypsin: interim results. Hum Gene Ther 22(10):1239–1247. doi:10.1089/hum. 2011.053

39. Mueller C, Chulay JD, Trapnell BC, Humphries M, Carey B, Sandhaus RA, McElvaney NG, Messina L, Tang Q, Rouhani FN, Campbell-Thompson M, Fu AD, Yachnis A, Knop DR, Ye GJ, Brantly M, Calcedo R, Somanathan S, Richman LP, Vonderheide RH, Hulme MA, Brusko TM, Wilson JM, Flotte TR (2013) Human Treg responses allow sustained recombinant adeno-associated virus-mediated transgene expression. J Clin Invest 123 (12):5310–5318. doi:10.1172/jci70314

40. Zern MA, Ozaki I, Duan L, Pomerantz R, Liu SL, Strayer DS (1999) A novel SV40-based vector successfully transduces and expresses an alpha 1-antitrypsin ribozyme in a human hepatoma-derived cell line. Gene Ther 6 (1):114–120. doi:10.1038/sj.gt.3300793

41. Cruz PE, Mueller C, Cossette TL, Golant A, Tang Q, Beattie SG, Brantly M, Campbell-Thompson M, Blomenkamp KS, Teckman JH, Flotte TR (2007) In vivo post-transcriptional gene silencing of alpha-1 antitrypsin by adeno-associated virus vectors expressing siRNA. Lab Investig 87 (9):893–902. doi:10.1038/labinvest. 3700629

42. Mueller C, Tang Q, Gruntman A, Blomenkamp K, Teckman J, Song L, Zamore PD, Flotte TR (2012) Sustained miRNA-mediated knockdown of mutant AAT with simultaneous augmentation of wild-type AAT has minimal effect on global liver miRNA profiles. Mol Ther 20(3):590–600. http://www.nature. com/mt/journal/v20/n3/suppinfo/ mt2011292s1.html

INDEX

Florie Borel and Christian Mueller (eds.), *Alpha-1 Antitrypsin Deficiency: Methods and Protocols*, Methods in Molecular Biology,
vol. 1639, DOI 10.1007/978-1-4939-7163-3, © Springer Science+Business Media LLC 2017